Ultrasound-Mediated Imaging of Soft Materials

Series on Advances in Optics, Photonics and Optoelectronics

SERIES EDITOR

Professor Rajpal S Sirohi Consultant Scientist

About the Editor

Rajpal S Sirohi is currently working as a faculty member in the Department of Physics, Alabama A&M University, Huntsville, Alabama (USA). Prior to this, he was a consultant scientist at the Indian Institute of Science Bangalore, and before that he was chair professor in the Department of Physics, Tezpur University, Assam. During 2000–11, he was academic administrator, being vice chancellor to a couple of universities and the director of the Indian Institute of Technology Delhi. He is the recipient of many international and national awards and the author of more than 400 papers. Dr Sirohi is involved with research concerning optical metrology, optical instrumentation, holography, and speckle phenomenon.

About the series

Optics, photonics and optoelectronics are enabling technologies in many branches of science, engineering, medicine and agriculture. These technologies have reshaped our outlook, our way of interaction with each other and brought people closer. They help us to understand many phenomena better and provide a deeper insight in the functioning of nature. Further, these technologies themselves are evolving at a rapid rate. Their applications encompass very large spatial scales from nanometers to astronomical and a very large temporal range from picoseconds to billions of years. The series on the advances on optics, photonics and optoelectronics aims at covering topics that are of interest to both academia and industry. Some of the topics that the books in the series will cover include bio-photonics and medical imaging, devices, electromagnetics, fiber optics, information storage, instrumentation, light sources, CCD and CMOS imagers, metamaterials, optical metrology, optical networks, photovoltaics, freeform optics and its evaluation, singular optics, cryptography and sensors.

About IOP ebooks

The authors are encouraged to take advantage of the features made possible by electronic publication to enhance the reader experience through the use of colour, animation and video, and incorporating supplementary files in their work.

Do you have an idea of a book you'd like to explore?

For further information and details of submitting book proposals see **iopscience.org/books** or contact Ashley Gasque on **Ashley.gasque@iop.org**.

Ultrasound-Mediated Imaging of Soft Materials

Ram Mohan Vasu and Debasish Roy
Indian Institute of Science, Bangalore

IOP Publishing, Bristol, UK

ISBN 978-0-7503-1750-4 (ebook)
ISBN 978-0-7503-1748-1 (print)
ISBN 978-0-7503-1749-8 (mobi)

DOI 10.1088/2053-2563/aae893

Version: 20181201

IOP Expanding Physics
ISSN 2053-2563 (online)
ISSN 2054-7315 (print)

British Library Cataloguing-in-Publication Data: A catalogue record for this book is available from the British Library.

Published by IOP Publishing, wholly owned by The Institute of Physics, London

IOP Publishing, Temple Circus, Temple Way, Bristol, BS1 6HG, UK

US Office: IOP Publishing, Inc., 190 North Independence Mall West, Suite 601, Philadelphia, PA 19106, USA

To our wives
Jessie and Runa

Contents

Preface

Ultrasound was introduced in near-infrared imaging of turbid objects to improve spatial resolution. Thus, diffuse optical tomography got a dual-model imaging tag, producing, to begin with, qualitative and then quantitative images with improved resolution afforded by tight focusing of the acoustic beam. The introduction of ultrasound forcing produced speckle modulation which was used as the measurement to recover absorption coefficient images. Ultrasound tagging, though helpful to improve spatial resolution, also gave the practitioner the dilemma of the many-to-one map, familiar to those dealing with inverse problems and also to the doctor who struggles to reach the right diagnosis from one symptom with a multitude of possible ailments. Here, the one measurement, the speckle modulation, has many causes, some from optical property changes and others owing to the dynamics—stochastic as well as deterministic. There is the ever present background Brownian dynamics the particles are subjected to, and presently the deterministic dynamics superimposed from the ultrasound forcing, and also dynamics from the mixing of the above two. Earlier, we were in the happy near-equilibrium situation for the medium, thermally as well as the dynamically. The temperature-induced perturbation of the system causes it to depart from equilibrium, only to return obeying well-known fluctuation dissipation theorems. Ultrasound upsets all these: earlier, the dynamics could be described as a stationary random process, but with the external force it is non-stationary; thermal equilibrium is lost with local entropy production and detailed balance laws are no longer valid; systems driven far away from equilibrium, in a dissipative medium, loses their time reversal symmetry and hence new fluctuation dissipation theorems valid away from equilibrium need to derived.

We are far from claiming that all of the new physics to understand the system response in its entirety at the ultrasound focal volume are described in this monograph. Our job here was only to give a 'voice in the wilderness' pointing out that there are issues to be tackled for a proper understanding of the behaviour of soft tissue-like material subject to thermal and acoustic loading. Should one spend time on these issues, just as we tried to in a modest way here, there are benefits one can reap. In the imaging context, some of them are (i) a full recovery of the elasticity tensor, (ii) measurement of rotational diffusion using light scattering, which can be very useful in studying mixing of linear and rotational particle diffusion, and also (iii) control of anomalous diffusion with the handle of ultrasound force. Our hope is that researchers in the field will be motivated to consider looking into these problems by the surface scratch we have given in this monograph.

It is our pleasure to thank all those who contributed to the work presented here: primarily, our students, Sriram Chandran, Saikat Sarkar, Dibbyan Mazumder and Mamatha Venugopal, who did all the experiments and computations as part of their PhD dissertations. We are especially thankful to Sriram and Saikat who helped us with last minute verification of some of the results and redrawing some of the figures.

We have also received enthusiastic help from our colleagues, A K Nandakumaran, in formulating the inverse problem, and Rajan Kanhirodan in setting up of the experiments. We also acknowledge the help we got from Institute of Physics Publishing Team, whose persistence helped us consider taking up this writing.

Author biographies

Ram Mohan Vasu

Ram Mohan Vasu finished his PhD in Physics from University of Aston in Birmingham in 1980. He joined The Indian Institute of Science, Bangalore, first as a post-doctoral Fellow and then as Assistant Professor in the Department of Instrumentation and Applied Physics. He became a full Professor in 2007 and retired from IISc in 2016. During his tenure at IISc he taught courses in Measurements, Signal Processing, Applied Optics, Imaging, Inverse Boundary Value Problems and short courses in Numerical Techniques. He has graduated altogether 34 students, 22 with PhD degrees and 12 with MS. All through, his interests were in imaging and signal processing with special emphasis on optics and tomography, having contributed to theoretical understanding of the inverse problem involved, experiments and also to a smaller degree instrument development. He has more than 100 research papers in areas mentioned above. Currently he is a consultant for Defence R&D Project granted to IISc.

Debasish Roy

Debasish Roy is currently a Professor at the Computational Mechanics Laboratory and the Convener of the newly established Centre of Excellence for Advanced Mechanics of Materials at the Indian Institute of Science, Bangalore. He obtained his PhD from the Indian Institute of Science, followed by post-doctoral research at the University of Innsbruck, Austria. Besides being a fellow of the Indian National Academy of Engineering, he has also held an Honorary Professorship in the School of Engineering, University of Aberdeen, a distinguished visiting fellowship of the Royal Academy of Engineering, London and a Visiting Professorship to the Department of Mechanical Engineering, Texas A&M University. His areas of research include computational mechanics of non-classical continua, stochastic dynamical systems and optimization/inverse problems. He has guided 23 PhD students, mentored several post-docs and published over 140 papers in journals of international repute. He has also delivered keynote/invited lectures at many international conferences and served on the editorial boards of several international journals of repute.

Chapter 1

Introduction

1.1 General introduction

Recovery of spectroscopic variation in optical absorption and scattering coefficients in soft-tissue organs, demonstrated by diffuse optical tomography (DOT), has given the physician quantitative information on certain important functional parameters. They are extremely useful for early diagnosis of pathology and also as molecular markers to map gene expression. However, DOT, owing to diffusion of photons and the consequent ill-posedness the inverse parameter estimation problem inherits, results in 'images' with poor spatial resolution. One way to counter the poor resolution is to make the inverse problem well-posed by localizing reconstruction to a chosen region—a region perhaps identified by the presence of another radiation and referred to as the region-of-interest (ROI). The second radiation employed in the context of DOT is ultrasound focused to a tight region which forms the ROI [1]. This hybrid method, wherein the changes (modulation) produced by ultrasound radiation in the ROI are carried by photons and measured, is christened ultrasound modulated optical tomography (abbreviated either UMOT or simply UOT) and the application is the recovery of optical contrast in the ROI. Spatial resolution is improved in as much as the ultrasound can be tightly focused to control the size of the ROI and a spatially discretized image is obtained by scanning the object with the ROI. A complementary opto-acoustic means to recover optical contrast at improved spatial resolution is photo-acoustic tomography (PAT); here an acoustic wave naturally arises through the photo-acoustic process when an object is probed by an optical pulse and carries signature on the optical absorption coefficient variation to the boundary for measurement. Good spatial resolution is achieved in shallow tissue layers, and in deeper regions the propagation of ultrasound through the intervening tissue renders the recovered 'sources' of poor resolution [2]. Since the initial pressure source which propagates as ultrasound waves is created by light absorption, the PAT is primarily developed to recover optical absorption contrast. The UMOT on the other hand is sensitive to both absorption and scattering contrasts and produces

doi:10.1088/2053-2563/aae893ch1

optical contrast images at all depths with spatial resolution limited by the size of the ultrasound focal volume. For quantitative recovery of optical contrast(s), partial differential equations that govern the propagation of acoustic and optical waves have to be inverted for their coefficients.

This monograph is concerned about UMOT, not for the recovery of optical contrasts as originally envisaged by its developers and almost all of its practitioners until now, but for mechanical properties, for example, the Young's modulus. Change in mechanical stiffness is a disease marker for malignancy in soft-tissue organs just as the optical absorption coefficient is, but with a sharper and larger overall variation and range. The measurement in UMOT, which is the modulation depth of the speckle formed by interference of coherent light interrogating the ROI, is strongly sensitive to elastic property changes in the material of the ROI. The reason is that the vibration characteristics of scattering centres change drastically with stiffness, which in turn control speckle modulation through the phase modulation of interrogating light. Importantly, the sensitivity of modulation to a material property like Young's modulus is much higher than to optical absorption or scattering coefficients [3]. The higher sensitivity can be attributed also to the cumulative accumulation of phase from scatterer displacements in the path of any optical beam influencing the speckle modulation (as noted in [4], coherent light in a diffusive medium helps to amplify movement). In consequence, UMOT offers a sensitive means to map, along with the optical contrast, the mechanical stiffness contrast as well. The road to mechanical property measurement of the host medium is through an understanding of the dynamics of the scattering centres therein. The property of light that responds readily to particle motion, or movement, is coherence and the measurement that we make is the decay of autocorrelation of light intensity (or amplitude), as is done in diffusing wave spectroscopy (DWS). Because of movement amplification available through multiple light scattering, we reap the benefit of sub-nm sensitivity to displacement measurement. We take advantage of this sensitivity in the ultrasound-assisted measurements reported here as well. Before we give an overview of topics discussed in this monograph, we first trace, very briefly, the historical development of UMOT in the context of optical contrast imaging.

1.2 UMOT in the context of optical contrast imaging

Initial attempts were to measure the speckle modulation depth (M) and use it to construct qualitative images of optical absorption coefficient ($\mu_a(\mathbf{r})$). In addition, change in speckle contrast was used to represent μ_s', the reduced scattering coefficient and through it the diffusion coefficient $\left(\kappa \equiv \frac{1}{3\mu_s'}\right)$ [5, 6]. Spatial variation of M, or change of speckle contrast when the ultrasound focal volume scanned the object, formed 'images' of $\mu_a(\mathbf{r})$(or $\mu_s'(\mathbf{r})$). Fractional intensity $\left(\frac{\Delta I}{I}\right)$ originating at the ROI affected by the local $\mu_a(\mathbf{r})$ can be obtained from the measured M. Possible early detection of malignancy based on contrast in $\mu_a(\mathbf{r})$, helped by the enhanced spatial resolution through tight ultrasound focusing, was the intended application.

Initial development saw progress in experimental techniques to detect the weak speckle intensity modulation from the background, noisy unmodulated light. The key to improving signal-to-noise ratio (SNR) is to capture a single speckle from a single-mode fibre as is done in DWS experiments, or use an optical spectrum analyser like a confocal Fabry–Perot interferometer [7] which allows real-time detection of signals. Other attempts to improve SNR included parallel detection of signal from a spatial distribution of speckles followed by averaging [8], detection through spectral hole burning [9], and photo-refractive crystals and polymers [10]. Temporal autocorrelation of signal from a single speckle also showed improvement in SNR and contrast of the absorption coefficient recovered [6]. In the experiments described in the present monograph, we measure the autocorrelation of the detected intensity from a single speckle, as is done in a standard DWS experiment.

Attempts to lift UMOT from its lowly qualitative imaging modality to one that can provide quantitative measure to the recovered parameters are of more recent origin [2, 11–13]. Of these, [11] puts forth both iterative and non-iterative methods to recover absorption as well as diffusion coefficients from purely non-interferometric measurement of photon flux from an object insonified by an acoustic wave. Ultrasound pressure-induced modulation of diffusion- and absorption coefficients are considered, and explicit schemes are devised to recover the optical parameters from measured fractional change in intensity, $\frac{\Delta I}{I}$, owing to the introduction of the acoustic wave. Since the measurement does not involve any temporal fluctuation of intensity, physics of the origin of ΔI can be explained based only on the elasto-optic effect; phase fluctuations and the coherent addition of scattered and non-scattered light need not be considered.

In [2, 12, 13], intensity fluctuations of the speckle formed by the coherent addition of the scattered photons that have intersected the ROI with those which have not, is the measurement. This is also referred to as the speckle modulation depth, or the modulated photon flux [12], or intensity of the modulated photons, isolated and measured in some cases, with the help of a Mach–Zehnder interferometer. A diffusion equation is used to model photon propagation through the object, and a perturbation equation (obtainable using the Frechet derivative of the first equation) derived from this, to model diffusion of modulated photon flux. Perturbation is caused by the ultrasound force in the ROI under which the material undergoes sinusoidal compression and relaxation producing similar density fluctuations seen by light as refractive index fluctuations (Δn). Ultrasound force also imparts periodic (can be approximated as sinusoidal) displacement to the scattering centres in the ROI. Light intercepting the ROI picks up a periodic modulation in phase owing to the combined effect of both Δn and displacement which leaves a periodic modulation in intensity to the speckle pattern resulting from the coherent addition of the scattered wavelets. This modulation is also affected by the dissipation present in the region where it had its origin: optical absorption represented by a non-zero $\mu_a(\mathbf{r})$. In addition, $\mu_s'(\mathbf{r})$ also affects the speckle modulation; for, the higher the number of scattering events, the larger will be the phase modulation picked up and the speckle modulation.

To compute the displacements suffered by the scattering centres, one needs an appropriate momentum-balance equation and the (visco) elastic properties of the material of the ROI are part of this equation. Visco-elastic properties are needed also to quantitatively study the ever present Brownian dynamics of the scattering centres. Owing to this stochastic movement, the periodic speckle fluctuation becomes noisy, and the modulation thereof decays, which can be observed in the temporal autocorrelation of the speckle intensity [14, 15]. This modulation decay represents the correlation decay suffered by light in the ROI and can be used to compute the mean-squared displacement (MSD) of the scattering centres within the ROI and also the local visco-elastic spectra, through the standard DWS theory. However, in the context of quantitative UMOT for optical contrast, the 'measurement' which is the strength of modulated photon flux inferred from the speckle power-spectral amplitude at the ultrasound frequency, ω_a, is compromised by the Brownian noise picked up by light. In fact, the power spectrum centred on ω_a gets broadened by superimposed random movement of the scattering centres resulting in measurement error which may adversely affect the optical contrast recovery. To the best of our knowledge, increased spread of the power-spectrum around ω_a and its effect on UMOT measurement is mentioned in just one of the publications [4]. The positive aspect of the increased bandwidth is extra information made available through the experiment. However, there is no demonstration of the effect of noise through assessment of optical contrast recovered through inversion of data. We mention in passing that equations (10) and (11) in [4] take into account the effect of Brownian motion of particles, modulation of refractive index and light absorption of the background material on modulation of the amplitude autocorrelation ($G(\mathbf{r}, \tau)$) of detected light, and its decay. One effect that is not accounted for is the periodic movement of scattering centres introduced by the ultrasound forcing.

1.2.1 History of quantitative optical contrast recovery with UMOT

As mentioned in the previous section, demonstration of increased spatial resolution in UMOT was for a long time limited to the creation of qualitative images of primarily $\mu_a(\mathbf{r})$ through the display of modulated photon flux collected, as the object is scanned by the focal volume of the ultrasound transducer. Validating or proving an inversion algorithm on measurement came almost as an afterthought in recent years and publications showing quantitative images from the measured data are not many. The earliest in this line is from Sefez and Lev [4], wherein a connection between optical properties and dynamics of the object medium and also the speckle correlation time is demonstrated. Since the periodic fluctuation in the speckle is introduced by the action of the ultrasound force at the ROI (arising out of mixing the modulated and unmodulated photons in the detector), local dynamics, e.g. local circulation in capillaries, and the fluctuating part of the intensity could be connected.

Relating the fluctuating part of intensity and inhomogeneity in optical contrast at the ROI was undertaken in [12, 13] and the experimental data was inverted to recover $\mu_a(\mathbf{r})$ in [12]. Analytical expressions connecting diffusion- and absorption coefficients to the photon flux, fluctuating and the background, are derived in [11]. However, in none of the above publications, contribution to the overall phase modulation from the dynamics induced by the ultrasound forcing, with its own share in the measured intensity fluctuation, is taken into consideration. As mentioned earlier, the ultrasound force leads to an almost sinusoidal modulation in $n(\mathbf{r})$ in the ultrasound focal volume. This periodic force sets in motion mechanical vibration of the scattering centres in the ROI. Since the major application of UMOT is imaging tissue with a very large bulk modulus (implying near-incompressibility), $\Delta n(\mathbf{r})$ should be close to zero. But the shear-stress induced dynamics in the ROI is not small and cannot be neglected. If one has to account for this dynamics contributing to the fluctuating part of the photon density through its own share in the phase fluctuation, the underlying mechanics must be brought in, for example, through momentum balance equations supplemented with a constitutive closure. Assuming the deformation to be small and the mechanical response linear elastic or visco-elastic, the mechanical properties of the material, such as shear- or Young's modulus, also enter the model describing the overall phase fluctuation creating the fluctuating intensity. This brings us to the main contribution of the work reported in this monograph: recovery of visco-elastic properties using UMOT with a spatial resolution afforded by the tight focusing of ultrasound in tissue. Recovery of mechanical stiffness in soft-tissue organs is equally important in diagnostic imaging as the optical contrast recovery; for malignancy enhances the mechanical stiffness of the tissue as it increases optical absorption through angiogenesis. Since the percentage change in, say, shear modulus as malignancy progresses far outstrips that of $\mu_a(\mathbf{r})$, mechanical property recovery can stand on its own merit as an important diagnostic tool, or at least as a complementary measurement to optical contrast. Before we conclude this chapter with a brief overview of the chapters to follow, we would like to make a few remarks on the existing work on quantitative UMOT for optical contrast reconstruction.

1.2.2 Overview of work on quantitative ultrasound-modulated optical tomography

Our first remark is concerning the omission of the ever present background Brownian dynamics of particles in the ROI in most publications (except in [4] where an exponential decay in the amplitude autocorrelation accounting for the Brownian motion is included in equations (10) and (11), for example). As indicated earlier, background Brownian dynamics decays the autocorrelation of detected light, and in the context of UMOT the modulation overriding the autocorrelation. When the power spectral amplitude at the ultrasound frequency is used to represent the modulated flux density and hence used as the measurement for the quantitative recovery of optical contrast [2] (both $\mu_a(\mathbf{r})$ and $\kappa(\mathbf{r})$), as already mentioned, the gradual decay in correlation can introduce error in measurement, affecting the accuracy of recovery. Also, the measurement in quantitative UMOT in examples such as [2, 13] is the power-spectral amplitude at ω_a, which represents the strength of the modulated photon flux. (We note in passing that correlation decay is the measurement for DWS from which visco-elastic spectra is recovered in micro-rheology [16]. As mentioned earlier, we extend DWS-based micro-rheology to inhomogeneous objects by the local reconstruction of visco-elastic spectra in the ROI via measurement of decay of the modulation introduced in $G(\mathbf{r}, \tau)$ (see chapter 2)). In [4], the measurement is the power-spectral width and is used to discriminate living tissue from *ex vivo* samples and phantoms. It is interesting to take note of the increased power-spectral width around ω_a in figure 9 of [4] when the living tissue is replaced by a liquid medium where the intensity of Brownian motion is high.

In order to account for all the sources contributing to fluctuation in $G(\mathbf{r}, \tau)$, denoted by $G^\delta(\mathbf{r}, \tau)$, one needs to widen the scope of the 'source term' (as given in [2, 12, 13]) in the partial differential equation (PDE) describing the propagation of $G^\delta(\mathbf{r}, \tau)$. Note that the property of light used in [2, 12, 13] is the fluctuating photon flux, which is $G^\delta(\mathbf{r}, 0)$. Since all the above publications deal with the same issue of how the presence of the ultrasound modulates the photon flux through induced phase modulation, the PDEs modelling the evolution of modulated autocorrelation in all of them are basically the same. Therefore, we will only have a look at the formulations as given in [2, 13] below.

The fluctuating part of the amplitude autocorrelation function as derived in [13] is proportional to $\alpha\int_\Omega \overline{G}(\sigma, x, 0)|p(x)|^2\overline{G}(x, \eta, 0)dx(1 - \cos\omega_a\tau)$. Here, $\overline{G}(\sigma, x, \tau)$ and $\overline{G}(x, \eta, \tau)$ are the probability density of photon path length weighted autocorrelation functions of light with source at σ and detector at x, and source at x and detector at η, respectively. Moreover, $|p(x)|^2$ is the squared amplitude of the acoustic pressure, α is a constant of proportionality and Ω is the support of the ROI. This is a convenient way of arriving at the strength of fluctuating photon flux as $\alpha\int_\Omega u(x)|p(x)|^2\overline{G}(x, \eta, 0)dx$ where $u(x)$ is used to represent $\overline{G}(\sigma, x, 0)$, the average intensity at x. This is the amplitude of modulation superimposed on the autocorrelation function through ultrasound forcing, which is used as data for the recovery of optical contrast [2, 12]. However, quite unfortunately $|p(x)|^2$ in the above expression does not capture the loss of coherence suffered by photons as they diffuse through the ROI with scattering particles that move because of ultrasound forcing and temperature (Brownian motion). In addition, there is dissipation because of local non-zero μ_a. Assuming that the optical pathlength travelled by a typical photon at time t at the jth scattering event is $n_j l_j$ which becomes $(n_j + \Delta n_j)(l_j + \Delta l_j)$ at time $(t + \tau)$, to a first order approximation the change in pathlength is $(n_j\Delta l_j + l_j\Delta n_j)$. Here, n_j is the refractive index of the medium in the neighbourhood of the jth scattering centre. Assuming the body to be nearly incompressible, the above change can be approximated by $n_j\Delta l_j$ which is denoted by the vector $n_j\Delta\mathbf{l}_j$. The corresponding phase change in time lag τ is $\Delta\varphi_j(\tau) \equiv \mathbf{q}_j \cdot \Delta\mathbf{l}_j$ where $\mathbf{q}_j$ is the scattering vector (the difference between incident- and scattered wave vectors). The overall contribution to phase fluctuation from light following a typical photon path s is $\Delta\varphi_s(\tau) = \sum_{j=1}^{n_s}\Delta\varphi_j$ where n_s is the number of scattering events along the path, within the ROI. (Here, we have neglected fluctuations due to Brownian motion inside and outside the ROI.) The contribution to $G_s^\delta(\mathbf{r}, \tau)$ (subscript s referring to the path of length s) from this phase fluctuation is $\exp\left[-\frac{1}{2}\left\langle \Delta\varphi_s^2 \right\rangle\right]$ [17]. Here, angular brackets indicate average over time. An expression for $\left\langle \varphi_s^2 \right\rangle$ is derived by many, either in the context of DWS [18] or UMOT [4]. From [4] we have

$$\left\langle \varphi_s^2 \right\rangle = \sum_{j=1}^{n_s} \frac{2}{3} k_0^2 \left\langle A^2(\mathbf{r}_j) \right\rangle \sin^2 \frac{\omega_a \tau}{2}. \tag{1.1}$$

Here, the sinusoidal radiation force from the focused ultrasound wave is assumed to produce in the ROI an oscillatory motion of amplitude $A(\mathbf{r}_j)$ at a typical scattering centre located at $\mathbf{r}_j$ and the number of scattering events a photon typically encounters along path of length s within the ROI is denoted by n_s which the diffusion approximation recognizes i.e. $n_s = \frac{s}{l^*}$ where l^* is the transport mean-free-path. (The particles have isotropic scattering only in the ideal case of perfect spheres in which case l^* becomes l_s, the scattering mean-free-path. However, particles have a residual non-sphericity, resulting in scattering anisotropy, which means that there are many scattering events within l^* which the diffusion model does not account for.) Moreover, $\left\langle A^2(\mathbf{r}_j) \right\rangle$ is the mean of $A^2(\mathbf{r}_j)$ evaluated over scattering centres within $(l^*)^3$ (or the volume of the ROI whichever is lower), and k_0 is the modulus of the propagation vector of light used. Using equation (1.1), we can arrive at $G_s^\delta(\mathbf{r}_s, \mathbf{r}_d \tau)$ as $G_s^\delta(\mathbf{r}_s, \mathbf{r}_d, \tau) = \exp\left[-\frac{2}{3} \sum_{j=1}^{n_s} k_0^2 \left\langle A^2(\mathbf{r}_j) \right\rangle \sin^2 \frac{\omega_a \tau}{2}\right]$. Summing up $G_s^\delta(\mathbf{r}_s, \mathbf{r}_d, \tau)$ over all the photon paths (again denoted by s, allowing a small abuse of notation), weighted by the probability density function for pathlength s, $p(\mathbf{r}_s, \mathbf{r}_d, s)$, gives

$$G^\delta(\mathbf{r}_s, \mathbf{r}_d \tau) = \int_s p(\mathbf{r}_s, \mathbf{r}_d, s) G_s^\delta(\mathbf{r}_s, \mathbf{r}_d, \tau) ds \tag{1.2}$$

Here, the integral is over all photon paths which have nonzero intercept over the support of the ROI; further, $\mathbf{r}_s$ and $\mathbf{r}_d$ are position vectors for the source and detector, respectively. The probability density function can be obtained by solving the photon diffusion equation with a delta-source at $\mathbf{r}_s$ making use of the background optical and mechanical properties of the object. Or in a general setting, since we are dealing with autocorrelation and its modulation, by solving the correlation diffusion problem [18] one can get $G_s(\mathbf{r}_s, \mathbf{r}_d, \tau)$ and the photon flux $I(\mathbf{r}_s, \mathbf{r}_d) = G_s(\mathbf{r}_s, \mathbf{r}_d, 0)$. Since $G_s(\mathbf{r}_s, \mathbf{r}_d, \tau)$ can be split as $G_s(\mathbf{r}_s, \mathbf{r}_d, \tau) = G_s(\mathbf{r}_s, \mathbf{r}_j, \tau) G_s(\mathbf{r}_j, \mathbf{r}_d, \tau)$, equation (1.2) can be rewritten as

$$\begin{aligned} G^\delta(\mathbf{r}_s, \mathbf{r}_d \tau) &= \int_s G_s(\mathbf{r}_s, \mathbf{r}_j, \tau) G_s(\mathbf{r}_j, \mathbf{r}_d, \tau) G_s^\delta(\mathbf{r}_s, \mathbf{r}_d, \tau) ds \\ &\simeq \int_s \sum_{j=1}^{n_s} G_s(\mathbf{r}_s, \mathbf{r}_j, \tau) G_s(\mathbf{r}_j, \mathbf{r}_d, \tau) \left\{ 1 - \frac{2}{3} k_0^2 \left\langle A^2(\mathbf{r}_j) \right\rangle \sin^2 \frac{\omega_a \tau}{2} \right\} ds \end{aligned} \tag{1.3}$$

Replacing the summation by an integral equation (1.3) becomes

$$G^{\delta}(\mathbf{r}_s, \mathbf{r}_d, \tau) = \int_s \int_{\Omega_{\mathrm{ROI}}} G_s(\mathbf{r}_s, \mathbf{r}, \tau) G_s(\mathbf{r}, \mathbf{r}_d, \tau) \left\{ 1 - \frac{2}{3} k_0^2 \langle A^2(\mathbf{r}) \rangle \sin^2 \frac{\omega_a \tau}{2} \right\} d\mathbf{r} ds \quad (1.4)$$

In arriving at equation (1.4), both Brownian motion and absorption inhomogeneity within the ROI are not considered. Brownian motion is a source of decay in correlation and the contribution to $G^{\delta}(\mathbf{r}_s, \mathbf{r}_d, \tau)$ is a multiplicative term proportional to $\exp\left[-n_s \frac{\tau}{\tau_0}\right]$. Here, τ_0 is the decay constant whose value depends on the 'intensity' of local Brownian motion in the ROI. Absorption within the ROI causes loss of photons contributing to $G^{\delta}(\mathbf{r}_s, \mathbf{r}_d, \tau)$ quantified by a multiplicative loss term of the type $\exp[-s\mu_a^{av}]$, where μ_a^{av} is the average absorption coefficient within the ROI intercepted by photon path length s therein. These two sources of additional decay, one τ-dependent and the other μ_a^{av}-dependent, allow us measurements from which local (i.e. ROI-specific) properties of the object can be recovered. For example, the τ-dependent exponential decay makes the modulation imposed on $G(\mathbf{r}_s, \mathbf{r}_d, \tau)$ to decay with τ. Measurement of modulation decay leads to the recovery of decay constant τ_0, from which the visco-elastic spectrum of the material can be recovered as is done routinely in DWS, with the difference that here the spectrum pertains just to the material of the ROI. This is the basis of local recovery of visco-elastic spectrum mentioned earlier. The other decay term which depends on the average absorption coefficient in the ROI also affects the modulation, presently causing an overall decrease because of dissipation within the ROI. However, the modulation is dependent on two other factors as well: local μ_s' (reduced scattering coefficient) and $\langle A^2(\mathbf{r}) \rangle$. Provided separation of the contribution to modulation from μ_a^{av} alone is possible, one can use modulation measurement to recover μ_a^{av} which is traditionally done in quantitative UMOT.

With the introduction of these two sources of modulation in autocorrelation our case is almost complete, except for the role played by the local optical diffusion coefficient, $\kappa(\mathbf{r})$, in creating the perturbation $G^{\delta}(\mathbf{r}_s, \mathbf{r}_d, \tau)$. In the expression for $G^{\delta}(\mathbf{r}_s, \mathbf{r}_d, \tau)$ in equation (1.4), $G_s(\mathbf{r}_s, \mathbf{r}_d, \tau) = G_s(\mathbf{r}_s, \mathbf{r}_j, \tau) G_s(\mathbf{r}_j, \mathbf{r}_d, \tau)$ appears on the right-hand side which is the Green's function for the correlation diffusion equation (CDE) [18] with explicit dependence on the average $\kappa(\mathbf{r})$ in the object. Local $\kappa(\mathbf{r})$ within the ROI affects $G^{\delta}(\mathbf{r}_s, \mathbf{r}_d, \tau)$ through s, which is dependent on the local μ_s'.

An expression for $G^{\delta}(\mathbf{r}_s, \mathbf{r}_d, \tau)$ can also be arrived at by solving the correlation perturbation equation obtained from the CDE, modelling the propagation of perturbation in autocorrelation introduced by the focused ultrasound wave at the ROI [19]. The perturbation equation is

$$\begin{aligned} &- \nabla \cdot \kappa \nabla G^{\delta}(\mathbf{r}_s, \mathbf{r}_d, \tau) \\ &+ \left(\mu_a^b + (1/3) k_0^2 \mu_s' [\langle \Delta r^2(\mathbf{r}, \tau) \rangle]_b - \frac{2}{3} k_0^2 \mu_s' [\langle A^2(\mathbf{r}) \rangle]_b \sin^2\left(\frac{\omega_a \tau}{2}\right) I_{\Omega} \right) G^{\delta}(\mathbf{r}_s, \mathbf{r}_d, \tau) \\ &= - \left\{ \mu_a^{in} + (1/3) k_0^2 \mu_s' [\langle \Delta r^2(\mathbf{r}, \tau) \rangle]_{in} - \frac{2}{3} k_0^2 \mu_s' [\langle A^2(\mathbf{r}) \rangle]_{in} \sin^2\left(\frac{\omega_a \tau}{2}\right) \right\} I_{\Omega} G(\mathbf{r}_s, \mathbf{r}_d, \tau) \end{aligned} \quad (1.5)$$

Here, we assume that the spatial variation in $\kappa(\mathbf{r})$ (and hence in $\mu'_s(\mathbf{r})$) is negligibly small. In addition, $\mu_a^{av} = \mu_a^b + \mu_a^{in}$, the mean-squared displacement of Brownian particles ($\langle \Delta r^2(\mathbf{r}, \tau)\rangle$) is written as $\langle \Delta r^2(\mathbf{r}, \tau)\rangle = [\langle \Delta r^2(\mathbf{r}, \tau)\rangle]_b + [\langle \Delta r^2(\mathbf{r}, \tau)\rangle]_{in}$, $\langle A^2(\mathbf{r}, \tau)\rangle = [\langle A^2(\mathbf{r}, \tau)\rangle]_b + [\langle A^2(\mathbf{r}, \tau)\rangle]_{in}$ and I_Ω is the indicator function for the ROI. Here the subscript (or, superscript) b signifies uniform background property and the subscript in denotes the inhomogeneous region where the focused ultrasound beam currently is. Once the background properties are ascertained, say, through correlation tomography, the measurement(s) from UMOT can be used to reconstruct properties in the ROI which are 'riding on' the background values. Towards this a mean-square error minimization problem can be set up and solved, as is done in [2]. However, the recovery of perturbation of properties contained in the 'source term' of equation (1.5) is a linear problem, which is easier to solve. A word about measurement: as in [2], it can be the strength of modulation in $G^\delta(\mathbf{r}_s, \mathbf{r}_d, \tau)$ available from its power-spectral intensity at ω_a measured at a number of detector locations. It can also be ascertained from the power spectrum of $G_F = G(\mathbf{r}_s, \mathbf{r}_d, \tau) + G^\delta(\mathbf{r}_s, \mathbf{r}_d, \tau)$ (which is the final measurement in an experiment) after removing the 'dc pedestal' from it at ω_a.

In [2, 12, 13] wherein quantitative recovery of μ_a and μ'_s from UMOT-based measurement of modulation depth is tried, both the dynamics induced by ultrasound forcing (deterministic) and temperature (stochastic) are overlooked as possible contributors to modulation. If the measured quantity, say m is dependent on μ_a, μ'_s and shear modulus G (the last property has a bearing on both A^2 and Δr^2), then the differential change in m when the ultrasound focal region moves to a new inhomogeneous region, cannot be correctly captured solely through partial derivatives with respect to optical properties; unless the object has no variation in G. At least in the application for detection of malignancy, tumour presents both an increase in μ_a and G; therefore, a numerically accurate recovery of say, μ_a requires recognition of dependence of m on G as well. As indicated earlier, in the present work, we are solely interested in the recovery of mechanical properties such as Young's and shear moduli. All the same, we need to account for the dependence of m on μ_a should we employ the formulation in [2] and hope to get quantitatively accurate results. Or we need to employ a measurement which does not depend on optical properties, which we did in part of the work reported here. In particular, the resonant modes of vibration of the ROI, which are not dependent on the optical properties, are used to reconstruct elastic properties of the material of the ROI.

The model which can describe the dynamics of the scattering centres within the ROI, acted upon by stochastic thermal forcing and the deterministic ultrasound forcing is a generalized Langevin equation (GLE). This is essentially a force-balance equation in which the optical properties of the medium do not enter. Therefore, to study the diffusive or anomalous, sub-diffusive movement of scattering particles (called the Brownian particles) in a complex visco-elastic environment or to extract properties of the surrounding medium through the dynamics of the Brownian particles, the GLE has been employed in [20, 21] (and the references cited therein). However, since in UMOT light is employed to probe and carry information on dynamics, optical properties of the medium enter the equations describing the propagation and interaction of light. Since Brownian particle dynamics affects primarily the coherence of light, measurements based on decay of coherence (as is done in DWS experiments) can be used to recover the dynamics without being affected by $\mu_a(\mathbf{r})$ variation in the medium. In ultrasound-assisted DWS if the decay of modulation depth can be used to infer the local dynamics represented by velocity autocorrelation or $\Delta r^2(\mathbf{r}, \tau)$ (the MSD), it will be noticed that the dynamics of a typical scattering centre in the ROI is not truly Brownian, but a non-Markovian stochastic process [22]. History dependence of the viscous drag force, excursion far away from thermal equilibrium owing to externally applied forces, influence of nonlocal interaction of the environment (i.e. the 'bath particles') on the system particle dynamics, enhanced micro-rotation suffered owing to the externally applied force all combine to give the movement of the scattering centres a sub-diffusive character. A GLE which incorporates a properly designed multiplicative noise in the restoration term can capture the nonlocal interaction caused by micro-structural length-scale effects excited by the micro-rotations [23]. As discussed in chapter 4, we arrive at the intensity and variance of the noise by considering the effect of the micro-rotations of the bath particles on the system particle under consideration, and thus capture almost exactly the experimentally observed behaviour of $\Delta r^2(\mathbf{r}, \tau)$ with lag time. An exponentially correlated random force, which maintains the thermal motion of the particles whose structure and intensity are also crucial in reproducing experimental behaviour of complex fluids, driven away from thermal equilibrium, also enters the GLE.

An alternative route to studying the dynamics without optical contrast getting in the way is to base the measurements on frequencies: as demonstrated in chapter 3 on resonant frequencies of the vibrating ROI. A lot of insight into the vibration of the ROI and the subsequent generation of an acoustic wave was gained after the discovery of vibro-acoustic (VA) effect and vibro-acoustic tomography [24]. Signatures of the ROI-centric shear modulus information in shear waves (which in the normal course would die down quickly on propagation) was carried by acoustic (i.e. pressure) waves, generated by the ultrasound forcing, for detection at the object boundary. Information on the resonant modes of vibration of the ROI measured from the VA-wave peaks have recently been used in the quantitative recovery of elastic modulus [25]. Detection of resonant modes of the ROI has also been demonstrated through UMOT experiments by looking for the frequency of ultrasound forcing at which the modulation in the measured intensity autocorrelation peaks [26]. This is the subject matter in the first part of chapter 4.

Micro-rotation, enhanced through external forcing of the particles through ultrasound, need not necessarily be carried by coherent light through phase modulation. This ability crucially depends on the size of the ROI. The ultrasound forcing creates both translational and rotational motion, leaving their contributions to the phase to be picked up by light. If, however, the particles are isotropic (or, equivalently, under diffusion approximation, undergo isotropic scattering with l^* as the scattering mean-free-path), light cannot carry phase changes introduced owing to rotation of particles. On the other hand, if the volume of the ROI is comparable to or less than $(l^*)^3$, geometric anisotropy in scattering particles helps the photons to pick up the phase changes introduced by rotation as well. This additional information has been helpful in extracting components of the elastic tensor of the material corresponding to micro-twist, which was not possible from phase fluctuations measured from translation alone. A smaller ROI also helps in the detection of speckle modulation, which is usually plagued by poor SNR, and deemed hardly detectable except through adopting special processing methods [8]. Since the ultrasound focal volume becomes a virtual source of modulated photons producing through coherent addition a speckle modulation at the detector, the spatial coherence of the source plays a crucial role in deciding the depth (or, visibility) of this modulation. Size of the source spoils spatial coherence. Therefore, it is to the advantage of the experimentalist to design the ROI to be small. The ROI can be made small by making it the intersection volume of the focal regions of two ultrasound transducers. They can be positioned so that the focal volumes cross at its thinnest middle region and the size can be manipulated through suitable apodization of the ultrasound transducers.

1.3 A brief overview of the work

As indicated earlier, our objective is to employ UMOT to study the dynamics of the ROI or, equivalently, the scattering centres in the ROI. Through this we would like to detect departure from the background Brownian dynamics when the focal volume encounters flow in a hidden capillary. Our objective also encompasses recovery of region-specific mechanical properties pertaining to the ROI from studying its dynamics. As a precursor to this, we first demonstrate in chapter 2, the measurement of ROI-specific $\Delta r^2(\mathbf{r}, \tau)$ of Brownian particles from correlation decay contributed from the ROI. Here, we also simulate the growth of $\Delta r^2(\mathbf{r}, \tau)$ from the GLE modelling the motion of a typical Brownian particle with non-local interactions captured with the help of Gaussian multiplicative noise. This local measurement capability is also used for the detection of non-Brownian dynamics whenever the ultrasound probe encounters regions contributing a nonzero drift term to the stochastic process describing the dynamics. One such occasion is the appearance of a capillary with laminar fluid flow within. We detect this by observing the increase in slope (from 1 to nearly 2) of the $\Delta r^2(\mathbf{r}, \tau) \cdot \text{vs} \cdot \tau$ plots. From the known ultrasound force with which the flow is stopped so that the slope becomes linear again, indicating pure Brownian motion, the flow volume is ascertained. Chapter 3 reports the results of an experiment where modulation depth is measured against ultrasound frequency. It is seen that there are frequencies at which the measured modulation peaked. These frequencies are identified as the resonant modes of the vibrating ROI which were used to recover the elastic properties of the material of the ROI. When the object is inhomogeneous in elastic properties, the variation in modulation when the ultrasound frequency is scanned is employed to tomographically recover the Young's modulus distribution within the ROI. Here, a second PDE connecting the amplitude of vibration of the scattering centres (observable as $\langle A^2(\mathbf{r})\rangle$) to Young's modulus distribution was required over and above the equation describing the photon diffusion through the object. In the next chapter, as the first part, the origin of VA-wave at the ROI is described and also its ability to carry shear modulus information pertaining to the ROI. This VA-wave detected at the boundary is used to measure the first natural frequency of vibration of the ROI. This information is inverted for the average Young's modulus of the material in the ROI. In the second part, recovery of orthotropic elastic tensor from animal tissue is demonstrated by identifying power-spectral peaks of the fluctuations in the plateau of $\Delta r^2(\mathbf{r}, \tau) \cdot \text{vs} \cdot \tau$ graph, obtained from an ultrasound-assisted DWS experiment, with the resonant modes of the ROI. The amplification of the usually weak higher order modes of vibration is explained using array-enhanced stochastic resonance. The measured resonant modes are inverted for elasticity tensor using a stochastic optimization procedure. In chapter 5, rotational diffusion within the ROI is experimentally demonstrated by measuring the mean square phase fluctuation and identifying the part corresponding to micro-rotation of the anisotropic scattering particles. It is further demonstrated that the contribution to mean-squared phase from micro-rotation can be inverted for the visco-elastic spectrum using the generalized Stokes–Einstein equation for rotational diffusion. For a mechanically isotropic material, it was verified to coincide with the visco-elastic spectrum obtained through translational diffusion. Average size of the scattering particles is obtained from the sum of the Prony series coefficients of the memory kernel used to model the history-dependent drag force and the shape anisotropy from the strength of the multiplicative noise used in the GLE modelling the dynamics. These parameters are extracted by matching the experimentally measured growth and fluctuations of the $\Delta r^2(\mathbf{r}, \tau) \cdot \text{vs} \cdot \tau$curve. Chapter 6 gives the concluding remarks. Apart from these chapters, there is one appendix giving details of the stochastic optimization scheme used for solving the elasticity parameters from the measured resonant frequencies and another, giving details of the derivation of the form and intensity of the multiplicative noise used in the GLE for accounting for nonlocal interaction among scattering particles.

References

[1] Sefez B G and Lev A 2002 Pulsed ultrasound-modulated light tomography *Opt. Lett.* **28** 1549–51

[2] Powell S, Arridge S R and Leung T S 2016 Gradient-based quantitative image reconstruction in ultrasound-modulated optical tomography: first harmonic measurement type in a linearized diffusion formulation *IEEE Trans. Med. Imag.* **35** 456–67

[3] Usha Devi C, Bharath Chandran R S, Vasu R M and Sood A K 2007 *J. Biomed. Opt.* **12** 034035

[4] Lev A and Sfez B 2003 *In vivo* demonstration of ultrasound-modulated light technique *J. Opt. Soc. Am.* **20** 2347–54

[5] Wang L-H and Zhao X 1997 Ultrasound-modulated optical tomography of absorbing objects buried in dense tissue-simulating turbid media *Appl. Opt.* **36** 7277–82

[6] Li H and Wang L V 2002 Autocorrelation of scattered laser light for ultrasound-modulated optical tomography in dense turbid media *Appl. Opt.* **41** 4739–42

[7] Sakadzic S and Wang L V 2004 High resolution ultrasound-modulated optical tomography in biological tissues *Opt. Lett.* **29** 2770–2
[8] Leveque S *et al* 1999 Ultrasound tagging of photon-paths in scattering media: parallel speckle modulation processing *Opt. Lett.* **24** 181–3
[9] Wang L V *et al* 2008 Pulsed ultrasound-modulated optical tomography through spectral hole-burning as a narrow-band spectral filter *Appl. Phys. Lett.* **93** 11111
[10] Wang L V *et al* 2013 High-sensitivity ultrasound-modulated optical tomography with a photo-refractive polymer *Opt. Lett.* **38** 899–901
[11] Bal G and Schotland J C 2010 Inverse scattering and acousto-optic imaging *Phys. Rev. Lett.* **104** 043902
[12] Bratchenia A *et al* 2011 Acousto-optic-assisted diffuse optical tomography *Opt. Lett.* **36** 1539–41
[13] Allmaras M and Bangerth W 2011 Reconstruction in ultrasound-modulated optical tomography *J. Inverse Ill-Posed Probl.* **19** 801–23
[14] Chandran R S *et al* 2014 Diffusing-wave spectroscopy in an inhomogeneous object: local visco-elastic spectra from ultrasound-assisted measurement of correlation decay arising from the ultrasound focal volume *Phys. Rev.* E **90** 012303
[15] Chandran R S *et al* 2015 Detection and estimation of capillary flow in a tissue-like object with ultrasound-assisted diffuse correlation tomography *J. Opt. Soc. Am.* A **34** 1888–97
[16] Harden J L and Viasnoff V 2001 Recent advances in DWS-based micro-rheology *Curr. Opin. Colloid Interface Sci.* **6** 438–45
[17] Wang L V 2001 Mechanism of ultrasonic modulation of multiply scattered coherent light: an analytical model *Phys. Rev. Lett.* **87** 043903
[18] Boas D A and Yodh A G 1997 Spatially varying dynamical properties of turbid media probed with diffusing temporal correlation *J. Opt. Soc. Am.* A **14** 192–214
[19] Mohanan K P 2017 *PhD Thesis* Indian Institute of Science, Bangalore ch 2
[20] Volkov V S and Leonov A I 1996 Non-Markovian Brownian motion in a visco-elastic fluid *J. Chem. Phys.* **104** 5922–31
[21] Mankin R, Laas K and Sauga A 2011 Generalized Langevin equation with multiplicative noise: temporal behaviour of the auotocorrelation functions *Phys. Rev.* E **83** 061131
[22] Sarkar S *et al* 2015 Internal noise-driven generalized Langevin equation for nonlocal continuum model *Phys. Rev.* E **92** 022150
[23] Roy D and Rao G V 2017 *Stochastic Dynamics, Filtering and Optimization* (Cambridge: Cambridge University Press)
[24] Fatami M and Greenleaf J F 1999 Vibro-acoustography: an imaging modality based on ultrasound-stimulated acoustic emission *Proc. Natl. Acad. Sci. USA* **96** 6603–8
[25] Mazumder D *et al* 2017 Quantitative vibro-acoustography of tissue-like objects by measurement of resonant modes *Phys. Med. Biol.* **62** 107–26
[26] Chandran R S *et al* 2011 Ultrasound modulated optical tomography: Young's modulus of the insonified region from measurement of natural frequency of vibration *Opt. Express* **19** 22837–50

Ram Mohan Vasu and Debasish Roy

Chapter 2

Localized measurement of dynamics and mechanical properties

2.1 Introduction

Diffusing-wave spectroscopy (DWS) is used to study temperature-driven Brownian motion of scattering centres in turbid, diffusive media, from which mechanical properties of the host medium are ascertained. In the initial attempts viscous fluids were the host media, which were later replaced by more complex visco-elastic fluids. It was the ingenuity of pioneers in this field such as Einstein [1] and Smoluchowsky [2] that kinetic theory was brought in to study what was essentially a thermodynamic problem, to establish the way Brownian particles disturbed by thermal stress return to equilibrium, following the so-called 'fluctuation–dissipation theorems'. For the Brownian particles, temporal evolution of MSD was measured, from which the viscosity (for liquids) or the complex visco-elastic spectrum (for the visco-elastic jelly) was extracted [3]. The measurement in DWS is intensity fluctuation, which is owing to phase fluctuation picked-up by photons in their journey through the object, suffering many scattering events. Because of this cumulative contribution to phase from many particles, individual displacements of the order of angstroms can be studied through the measured phase fluctuations. Therefore, DWS was able to throw light on the microscopic origin of visco-elastic properties of jelly-like soft materials [4]. Since the measurements are averages over photon paths, the extracted MSD growth over time delay pertains to the entire object and therefore, the recovered mechanical properties are averages over the object studied. There were attempts to extend DWS to inhomogeneous objects, the most fruitful of these involve tomographic recovery of dynamic properties from boundary measurements of amplitude autocorrelation ($g_1(\tau)$) of photons exiting the object following diffusive paths. This inhomogeneous property recovery was christened diffuse correlation tomography (DCT) [5, 6]. In addition, there were also attempts to recover changes in dynamics,

doi:10.1088/2053-2563/aae893ch2

such as flow velocity through embedded capillaries, by analyzing decay in $g_1(\tau)$ [7] without bringing in tomography.

This chapter reports non-tomographic recovery of mechanical properties, exactly as is done in DWS, from an object with an inhomogeneous distribution of these properties. In this, we delineate a region in the object with the help of a focused ultrasound beam (earlier designated the ROI) and measure the correlation decay caused by the Brownian scattering particles in the ROI. Once such measurement is available, the recipe of inversion as in DWS is followed recovering the average properties within the ROI. When the cause of inhomogeneity is fluid flow in an embedded capillary tube, qualitative (as well as quantitative) detection of the same is also demonstrated through a similar ultrasound-assisted technique.

The first task in the demonstration of such 'imaging' in an inhomogeneous medium is to establish a model which describes the mixed dynamics of the particles in the ROI. This is done in section 2.2, using a generalized Langevin equation (GLE) (a stochastic force-balance equation) to model the predominantly translational degree of freedom of a representative system particle which stands for the continuum, the ROI. Such a theoretical model, describing the dynamics accurately on the one hand, and easy to implement numerically on the other, is required to help us validate the experimental measurements. The setting-up and solution of the GLE through a stochastic projection technique is also described in section 2.2. (The theoretical basis of the second experiment is also given in sections 2.3.1 and 2.4.2.) The next task is description of the measurement procedure to extract 'local' correlation decay in an earmarked region. This is demonstrated in section 2.3 along with the signal processing requirements to achieve such measurements. The experiments, first in an inhomogeneous jelly-like material made of (poly) vinyl alcohol (PVA) and the other in a PVA slab with a hidden capillary tube containing a laminar flow of water with scattering properties similar to the background slab, are given in section 2.4. Results are discussed in sections 2.4.1.2 and 2.4.2.2. The final remarks are in section 2.5.

2.2 Theory

2.2.1 Model describing the dynamics of the ROI

Perhaps there were no attempts in the past to extend DWS-based recovery of visco-elastic spectra to objects with inhomogeneities, because of the inability to measure correlation decay pertaining only to Brownian motion from a selected region. The only exception was a recent publication involving some of us [8], wherein a focused ultrasound beam was used to superimpose a deterministic dynamics in a selected region, and the simultaneous presence of the two dynamics paves the way for measuring Brownian motion within the selected region. The principle of extraction of the stochastic component of the dynamics from the signatures of the deterministic one, which is confined to an *a priori* selected ultrasound focal region, is based on the mixing of these dynamics. This is easily seen in the detected autocorrelation of light, in which the ultrasound-induced sinusoidal movement causes a modulation, which suffers decay because of the Brownian motion pertaining only to the region. Before we describe the signal processing required for extracting this data from the measured autocorrelation of light, we model the dynamics of the ROI acted upon by both thermal- and ultrasound-induced stress.

How the deterministic dynamics 'mixes' with the stochastic one, and how it is reflected in the measured autocorrelation of light are not properly dealt with in literature. Since local properties ($\mu_a(\mathbf{r})$ in UMOT and $E(\mathbf{r})$ in ultrasound-modulated elastography, UMOE) are extracted from this mixed signal, for quantitative accuracy, it is essential that we have a physically accurate model for the dynamics, and a signal processing procedure which can extract either the deterministic or stochastic part of the dynamics from the measured signal. In the following, our objective is to model the dynamics of the ROI. When the space-time length scale of the forcing or induced deformation becomes comparable to the material length scale, the system response at the continuum level is greatly influenced by the material microstructure. Examples of such materials are polymers, granular solids, and more importantly in our case, various types of tissue. In such cases, a strictly classical continuum hypothesis-based force-balance equation accounting only for local interactions fails to capture the material response. Nonlocal modelling which takes into account long-range interaction of micro-structural constituents by introducing length-scale effects in the constitutive equations is seen to capture the deformation in closer conformity to experimental observations. Some of the successful nonlocal models use micro-polar, micro-morphic formulations or gradient theories as the basis [9, 10]. In the present case of modelling the response of the ROI in a tissue-like material, we follow a different route and consider it to be made-up of a collection of harmonic oscillators. We capture the dynamics of the entire object through that of a representative system particle (harmonic oscillator) which is assumed to have a predominant, single (translational) degree of freedom (DOF). The long-range interactions are modelled through the influence of the collection of all other particles on the movement of the system particle. The evolution of the translational DOF of the representative particle affected also by the presence of all other particles (the so-called bath particles) is modelled eminently by the GLE. Thus an expedient way to replace the infinite-dimensional continuum is provided by the GLE modelling a system particle. Despite this, the standard form of GLE does not capture nonlocal interactions represented by the influence of the system of bath particles. Our motivation to modify the GLE so that length-scale information, the backbone to take in nonlocal interactions, can be incorporated is from the fluctuations in the plateau region of the measured MSD versus time curve [11] from a PVA slab (see also figure 2.5 in section 2.4.1.2). It is noticed that when ultrasound forcing is on, there is significant fluctuation in the plateau, which is absent when the fore is withdrawn. These fluctuations could not be modelled using the standard form of GLE because they arise from nonlocal micro-structural interactions.

Before we move on to modification of the GLE to incorporate micro-structural interactions, a word about the mixing of dynamics (produced by thermal stress and ultrasound forcing) as observed in the fluctuations in figure 2.5. One of the assumptions normally used in UMOT experiments is that the externally applied sinusoidal forcing of the ultrasound is small compared to thermal forcing. In other words, the dimensionless Peclet number, which is a measure of the relative strength of the externally applied force vis-a-vis the thermal forces, is kept small. With this, a fractional Brownian motion (fBm) describes the overall dynamics of scattering particle with a 'small' deterministic sinusoidal perturbation superimposed. Solution to the modified GLE gives this 'mixed' overall dynamics. The essentially low frequency response of the ROI to the sinusoidal forcing influences and, in turn, is influenced by, the comparatively high frequency response of thermal forces. For example, as further discussed in chapter 4, fluctuations in the MSD versus time plateau contain information on the natural frequencies of ROI excited by ultrasound forcing, but multiplied by the characteristic $\frac{1}{\omega^2}$ power-spectrum of the background Brownian motion. The presence of external forcing can influence the Brownian motion so that the early rise of MSD with τ follows τ^{α} with $0 < \alpha < 1$. However, as shown in [8], the influence of external forcing on MSD is cumulative, dependent on an integral up to current time, of a term dependent on square of amplitude of scattering particle vibration (A^2). Therefore, the effect of external forcing on MSD will be noticeable only beyond $\tau \approx O\left(\frac{1}{A^2}\right)$. Since the Peclet number is kept small, the initial rise of the MSD with time is unaffected by the application of external forcing. As further discussed in section 2.4.1, this invariance of the rise time is important for the accurate extraction of local visco-elastic spectra from an inhomogeneous object.

2.2.2 Modified generalized Langevin equation

The aim here is to derive a GLE that carries micro-structural information which would facilitate nonlocal interaction; or, in other words, take into account the effect of bath particles on the dynamics of the representative system particle. For doing this, we follow a micro-polar continuum formulation. The underlying principle used is that the strain operator associated with micro-rotations does not commute with its translational counterpart, thus introducing an uncertainty relation involving them. This gives rise to a randomness, a measure of which can be characterized through a noise. This is seen to be an internal noise, multiplicative in nature modulating the restoration kernel in the GLE. Through this internal noise term, evolution of the micro-structural interaction, as manifested in the macro-scale, is characterized. In addition, we have the usual additive noise as in a standard GLE, whose correlation structure is delineated through the second fluctuation dissipation theorem of Kubo [12]. Experimental observations of steady-state fluctuations from an inhomogeneous PVA slab with ultrasound modulation are accurately reproduced by numerical solution of the modified GLE.

Here, in this subsection, we give the summary of the route taken to arrive at the GLE with the details deferred to appendix A. The cornerstone of our strategy is based on the premise that micropolar theory helps to have a phenomenological approach to include length-scale effects, which in turn originate from material micro-structure, in the continuum response of objects to external forcing. In the micropolar model every material point is endowed with six DOFs, three translational for the macro-element and three rotational for the micro-element. In addition, the existence of couple traction vectors along with the usual force traction vectors of classical continuum is assumed in order to describe force transfer across interfaces. As the first task in deriving the equation of motion of the system particle with predominantly translational DOF we derive the Hamiltonian, H, of the discrete representation of the object which is the ROI. As noted earlier, the discrete object has a system particle surrounded by a set of bath particles. Apart from the system DOF all other DOFs appearing in H are referred to as bath DOFs and denote H by $H = H_s + H_b$. Since our interest is in deriving the equation of motion for the system DOF, all the micro-rotational DOFs appearing in it need to be eliminated. This is achieved by treating each micro-rotational DOF as a zero-mean random variable with all the DOFs evolving as random processes contributing altogether a noise term to the GLE describing the evolution of the predominant translational DOF. The presence of such a noise term quite naturally brings in non-determinism, an essential aspect of non-local interactions. In addition to this multiplicative noise, there is an additive noise arising out of variations in initial condition owing to thermal fluctuations. The final form of the GLE, (see equation (A.10) in appendix A) in its usual second order form, is given by

$$m\ddot{u} + ku + \int_0^t \eta(s-t)\dot{u}(s)ds = F(t) + \xi(t) + \int_0^t W_s^t(s-t)u(s)ds \qquad (2.1)$$

Here, u is the predominant component of displacement of the system particle, η is the damping memory kernel, $\xi(t)$ is the additive noise which also represents the thermally induced random forcing and $F(t)$ is the deterministic external forcing. Moreover W_s^t is the multiplicative noise arising out of the overall contribution of the micro-rotational DOFs of the set of bath particles.

In the next sub-section we describe how equation (2.1) is solved.

2.2.3 Solution of the Langevin equation

One is tempted to follow Monte Carlo simulation to integrate the GLE of equation (2.1); however, this route would lead to difficulties in making a choice of Δt, the time step-size, for the simulated paths are sensitively dependent on the choice of the step size. On the other hand, as shown in [8], a direct integration of the GLE for $x(t)$ is possible. Prior to doing this integration, we project the GLE to a higher dimensional system of stochastic differential equations (SDEs). Through this, the solution is ensured to have Markovian property, whilst the original solution of the GLE clearly is not a Markovian stochastic process. The random forcing term $\xi(t)$ is first decomposed into a sum of independent white-noise processes, i.e. $\xi(t) = \sum_{j=0}^{N-1} \vartheta_j(t)$ with $\langle \vartheta_j(s)\vartheta_j(t) \rangle = k_B T \exp[-v_j|t-s|]$. Invoking the FDT we can write the memory kernel as

$$\eta(t-s) = \sum_{j=0}^{N-1} \eta_j \exp[-v_j|t-s|] \qquad (2.2)$$

Here, $v_j = \frac{v_0}{d^j}$ is the inverse autocorrelation time of the jth filtered white-noise component and d a dilation parameter. In addition, v_0 is the high-frequency cut-off of $\xi(t)$, $\eta_j = \frac{\eta_\alpha v_0}{\Gamma(1-\alpha)d^j}C(d)$, the jth weight coefficient in the expansion of η, $\Gamma(.)$ the gamma function and $C(d)$ a constant.

We first reiterate that each $\vartheta_j(t)$ in the expansion of the forcing function is a solution to the first-order SDE

$$d\vartheta_j = -v_j\vartheta_j dt + \sqrt{2\eta_j v_j k_B T}\, dW_j(t) \tag{2.3}$$

Here $\{W_j(t)\}$ is a family of delta-correlated N-independent standard Brownian processes. Using this, it is easy to have a Markovian representation of the GLE as a system of $(N + 2)$ dimensional system of SDEs:

$$dx(t) = v(t)dt \tag{2.4a}$$

$$dv(t) = \left(\frac{-\omega^2}{m}\right)xdt + (1/m)\sum_{j=0}^{N-1} u_j(t)dt \tag{2.4b}$$

$$du_j(t) = [-\eta_j v_j - v_j u_j(t)]dt + \sqrt{2v_j\eta_j k_B T}\, dW_j(t) \tag{2.4c}$$

In the ROI there is sinusoidal forcing from the ultrasound operating at ω_f. With this, equation (2.4b) becomes

$$dv(t) = \left(\frac{-\omega^2}{m}\right)xdt + (A/m)\sin(\omega_f t)dt + (1/m)\sum_{j=0}^{N-1} u_j(t)dt \tag{2.5}$$

One can presently write the general solution of the GLE as

$$\mathbf{X}_t = \mathbf{\Phi}_t\mathbf{X}_0 + \mathbf{\Phi}_t\int_0^t \mathbf{\Phi}_s^{-1}\mathbf{F}_s ds + \mathbf{\Phi}_t\int_0^t \mathbf{\Phi}_s^{-1}\mathbf{H}_s d\mathbf{W}_s \tag{2.6}$$

Here, $\mathbf{X}_t = \{x_t, v_t, u_0, \ldots . u_{N-1}\}^T$ is the solution vector, $\mathbf{\Phi}_t$ is the fundamental solution matrix (FSM) and $\mathbf{F}_t$ is the deterministic force vector containing the sinusoidal forcing term with amplitude, A, as the only non-zero entry, $\mathbf{H}_t$ the diffusion coefficient matrix and $\mathbf{W}_t$ a suitable zero-padded vector of pure Brownian components.

A stochastic Heun scheme is used to integrate the coupled SDEs of equation (2.4). For details of the numerical scheme see [13]. The integration will provide us with an ensemble of paths, within a Monte Carlo framework, from which the weights for computing the averages $\langle x(t)^2\rangle$ are calculated. From $\langle x(t)^2\rangle$ the MSD $\langle\Delta x(t)^2\rangle \equiv \langle x(t)^2 - x(0)^2\rangle$ is computed. (In this way we use both t and τ to represent both time and time lag.) This is done by solving the GLE without the sinusoidal forcing, i.e. equation (2.1) without $F(t)$ on the right-hand side. The numerical values of parameters such as α, v_0, d, N and $C(d)$ are carefully chosen depending on the mechanical properties of the object. The results of simulation are given along with their experimental counterparts in section 2.4.1.2 after describing the experiments.

Application of a Monte Carlo framework did not succeed in providing a satisfactory solution of the GLE with the external forcing term. The reason is, as mentioned earlier, the high sensitivity of simulated paths to the choice of time step-size, Δt. We have verified through simulations that the parameters v_j and η_j have a large range of variation as j increases, making $\mathbf{\Phi}_t$ ill-conditioned. The error accumulated in integration to arrive at the drift term will have a deleterious effect on the final computational results, especially so taking into account also $\mathbf{F}_t$ the deterministic forcing term for larger ω_fs. The tissue-like object such as the PVA phantom used in the experiments reported in section 2.4.1, $\omega_f \approx 1$ kHz or greater is a typical case, and the computed drift terms are quite erroneous. Moreover, as the amplitude of forcing becomes larger, there is a greater possibility of external force influencing such system parameters as $\eta(t - s)$ and ω^2 which would render Kubo's second fluctuation dissipation theorem untenable in the present case. Then the assumption of small Peclet number is no longer valid, the basis for arriving at the correlation structure of additive stochastic forcing, and the present model is rendered inadequate in accurately capturing the dynamics.

Since we have difficulty in simulating the response of the system particle under the influence of both stochastic (internal) and sinusoidal (externally applied) forcing by numerically integrating the GLE we resort to an alternative approach based on change of measures, the details of which are available in the appendix of [8]. There, an analytical expression for the growth of MSD is derived (equation (A.14) of [8]). Drawing upon the derivation in [8] our objective primarily is to show that the transient in the MSD versus τ curve, i.e. during the initial period, is unaffected through the introduction of the external force. Again, from equation (A.14) of [8] one can verify that the last term on its right-hand side is the contribution of the external forcing to the MSD. It has a monotonically increasing contribution with time, since it is an integral with a strictly positive integrand. However, when τ is small, because of the presence of the small A^2 term its contribution is negligible. One needs to consider the effect of this term only from $\tau \approx O(1/A^2)$. The other advantage of this approach of integration is that errors in modelling of the dynamics are accounted for in a weak stochastic sense. When τ becomes large MSD attains a plateau, which is a steady state of the solution of equation (A.14) in the sense that $\pi_\tau(x^2) = \pi_{\tau+\Gamma}(x^2)$ where $\Gamma = \dfrac{\pi}{\omega_f}$ gives the 'time period'. The introduction of the external forcing has the effect of pushing up the plateau high compared to the one driven only by thermal stress, as seen in figures 2.4 and 2.5.

It is seen from the above derivation of the GLE and its solution that the introduction of external forcing has only affected the behaviour of the steady-state fluctuations in MSD in its plateau with almost no effect on its early rise to this plateau. This observation is important, for we extract the local visco-elastic spectrum from the rise of the MSD, which is in no way affected by the introduction of the ultrasound forcing. Therefore, the ultrasound-assisted extraction of these properties in an inhomogeneous medium (which is an average of the properties within the ROI) should be identical to those extracted from a homogenous object with properties of the material in the ROI.

2.3 Extraction of correlation decay introduced by Brownian particles in the ROI

Introduction of ultrasound forcing causes a sinusoidal modulation of scattering centres, predominantly within the ROI (outside the focal region the force is assumed quite small and the movement caused negligible), which is seen in the detected light as a sinusoidal phase modulation. This causes a periodic intensity modulation in the detected speckle intensity. Since the scattering centres themselves are also undergoing a stochastic movement (Brownian motion) there is a random phase fluctuation overriding the periodic phase variation. Therefore, the detected intensity modulation acquires decay with respect to time. This is similar to the decay in $g_1(\tau)$ which is owing to the Brownian motion 'seen' by the photons on their transport through the entire object. However, the decay in modulation is owing to the Brownian motion experienced inside the ROI. If we have the means to measure this decay in modulation ($M(\tau)$) then, as is routinely done in DWS which uses the decay of $g_1(\tau)$ for computing the average visco-elastic spectrum, we can compute the visco-elastic spectrum of the material within the ROI from this decay. In the next subsection we describe the signal processing required to arrive at this decay.

2.3.1 Computing the decay in modulation on amplitude autocorrelation

Our first observation is that the introduction of the external sinusoidal forcing and the consequent oscillatory dynamics makes the position vector of a representative Brownian particle non-stationary. Light in its traverse through a collection of such Brownian particles picks up a phase fluctuation which is non-stationary. This renders the light amplitude (or, intensity) a non-stationary random process. In consequence, we cannot apply the usual concepts of autocorrelation and power spectrum while processing the light amplitude or intensity to arrive at $M(\tau)$. In what follows we employ the concepts of evolutionary autocorrelation and power spectrum [14] to recover $M(\tau)$ at a discrete set of τ -values.

2.3.1.1 Evolutionary autocorrelation and power spectrum

In order to process signals which are random processes and non-stationary, we extend the concepts of autocorrelation and its Fourier transform, the power-spectrum, used for dealing with stationary signals. Towards this, the concepts of evolutionary autocorrelation and evolutionary power-spectra are introduced [14]. Consider a signal $x(t)$ which is a wide sense stationary (WSS) stochastic process. We can represent $x(t)$ through its uncorrelated Fourier spectral components $X(\omega)$ through

$$x(t) = \frac{1}{2\pi}\int_{-\infty}^{\infty} X(\omega)\exp[j\omega t]d\omega \tag{2.7}$$

Denoting by $r(\tau)$ the autocorrelation of $x(t)$ ($r(\tau) = E[x(t)x^*(t + \tau)]$) where E is the expectation operator, because of the uncorrelatedness of $X(\omega)$ we can write

$$E\{X(\omega_1)X^*(\omega_2)\} = R(\omega)\delta(\omega_1 - \omega_2). \tag{2.8}$$

Here, $R(\omega)$ is the Fourier transform of $r(\tau)$. This way, $R(\omega)$ gives power distribution versus frequency in the signal, and is therefore called the power-spectral density. When $x(t)$ is noisy, such as the measured light intensity or amplitude, we can write $X(\omega)$ as $X(\omega) = X_d(\omega)\sigma(\omega)$ where $\sigma(\omega)$ is white-noise and $X_d(\omega)$ represents the deterministic signal component in $x(t)$. Since white-noise is uncorrelated it is easily seen that

$$E\{X(\omega_1)X^*(\omega_2)\} = |X_d(\omega_1)|^2\delta(\omega_1 - \omega_2) \tag{2.9}$$

Therefore, $R(\omega)$ represents the deterministic power-spectral distribution in $x(t)$. However, when $x(t)$ is non-stationary, and $X(\omega)$ uncorrelated, representation of $x(t)$ through equation (2.7) is not valid. For certain non-stationary processes a representation using correlated coefficients, $\{X(\omega)\}$, has been suggested which generalizes $X(\omega)$ to a function of two variables: $X(\omega_1, \omega_2)$ with one of them tracking correlatedness. When dealing with a single realization of $x(t)$, as with an ergodic process, it becomes quite cumbersome to assess correlation without binding assumptions on the smoothness of $X(\omega_1, \omega_2)$ [15]. Therefore, an alternative based on time-varying power spectrum is introduced in [14].

In this representation the finite time interval over which signal is measured, say, $[0, T]$ is split into sub-intervals $(t, t + dt)$. We select dt such that in this sub-interval one may be allowed to assume the signal stationary, and look for a power-spectral distribution for a stationary signal. By varying t one can now have a power-spectral distribution evolving with time. In other words, one has an 'evolutionary' power spectrum. We now proceed to determine the interval dt over which we can assume stationarity for the signal. Towards this, we generalize the representation of signal given in equation (2.7) to include also non-stationary signals: i.e.

$$x(t) = \frac{1}{2\pi}\int_{-\infty}^{\infty} X(\omega)B_t(\omega)d\omega \tag{2.10}$$

Here another basis function $\{B_t(\omega)\}$ indexed by time takes the place of the Fourier kernel $\exp[j\omega t]$. The real variable ω is a generalization of frequency, which can be regarded as just frequency in this case. With $\{B_t(\omega)\}$ properly selected, $X(\omega)$ retains orthogonality as in stationary signal representation. One note of caution is that $X(\omega)$, and the signal representation, is not unique but tied to the selection of the family $\{B_t(\omega)\}$. In order to retain the notion of frequency with respect to non-stationary signals $B_t(\omega)$s are chosen as slowly time-varying envelope functions of the type $B_t(\omega) = \psi_t(\omega)\exp[j\varphi(\omega)t]$. Here, $\varphi(\omega)$ is the modulation frequency, and the slowly-varying envelope $\psi_t(\omega)$ is centred around $\omega = 0$ and has a representation

$$\psi_t(\omega) = \int_{-\infty}^{\infty} \tilde{\psi}_\omega(f)\exp[jft]df \tag{2.11}$$

Using this, we can represent a non-stationary signal by

$$x(t) = \frac{1}{2\pi}\int_{-\infty}^{\infty} X(\omega)\psi_t(\omega)\exp[j\omega t]d\omega \tag{2.12}$$

From equation (2.12) it is seen that $x(t)$ is expanded using sinusoids $\{\exp[j\omega t]\}$ weighted by the family of slowly-varying (sinusoidally modulated) envelope functions $\{\psi_t(\omega)\}$. Following the notion of power spectrum of stationary signals, we define for a non-stationary signal $x(t)$ evolutionary power spectrum:

$$X_t(\omega) = E[|X(\omega)|^2]|\psi_t(\omega)|^2 \tag{2.13}$$

where the ensemble average is taken with respect to the family of envelope functions, and $X_t(\omega)$ itself is defined with respect to the chosen family of functions. Therefore, for a given $x(t)$ the evolutionary power spectrum is not unique.

Since $\psi_t(\omega)$ is slowly varying in time, its temporal Fourier transform will be concentrated around $\omega = 0$. As discussed in [14] a measure of this concentration can be spectral width when the magnitude spectrum drops by $(1/e)$ of its zero-frequency value. Denoting this width by C_F we obtain a time interval dt over which the signal can be treated as stationary as $dt_S = \frac{1}{Sup\{C_F\}}$. With dt_S arrived at this way, we compute the evolutionary autocorrelation as $r(t, t + dt_S) = E[x(t)x^*(t + dt_S)]$, where the ensemble average is computed by time-averaging $x(t)x^*(t + dt_S)$ over the window centred at t. $X_t(\omega)$ is obtained by Fourier transforming $r(t, t + dt_S)$ with respect to time within the window. To process the experimental data here (i.e. the light intensity) we use a simple rectangular window.

2.3.1.2 Processing of amplitude autocorrelation for recovering modulation decay

Here we give some details of processing the measured $g_1(\tau)$ with modulation to recover the decay of modulation ($M(\tau)$). Measurement of intensity autocorrelation from which $g_1(\tau)$ is arrived at is given in the experimental section, section 2.4.1.1. Further details of experiments to provide ergodicity to an otherwise non-ergodic, visco-elastic object, and details of data processing required to extract rise of MSD with time from the decay of $M(\tau)$ are also given in section 2.4.1.1. In a standard DWS experiment, $g_2(\tau)$ is computed from an ensemble of intensity measurements using a hard-wired autocorrelator. In such cases where $g_2(\tau)$ has to be computed over a large range of τ, the routine in the autocorrelator uses the so-called 'multi-tau' scheme wherein the sampling is non-uniform (with a quasi-logarithmic spacing) and τ ranges from μs to tens of seconds [16, 17]. But presently, in order to extract $M(\tau)$ one needs uniform sampling at a rate set by the frequency of the ultrasound forcing (at the high end) and above the Nyquist rate set by the bandwidth of signal we want to recover (at the lower end). Hence, the photon-mode scheme, which allows uniform sampling, is used. The shortest sample time is set by the hardwired autocorrelator used in the experiments in section 2.4.1, which is 12.5 ns for the one used in the experiments: DAC Flex 021D from www.correlator.com, Bridgewater, NJ 08807. The intensity data collected in the 'photon mode' is auto-correlated using MATLAB 1-D autocorrelation routine [18] to arrive at $g_1(\tau)$. The rectangular window is slid across the measured $g_1(\tau)$ at a step-size $\Delta\tau$ determined earlier through the above mentioned criteria. At each position of the window denoted by $\tau = \tau_i$, $M(\tau_i)$ is evaluated as the modulus of the Fourier transform of the signal inside the window at the modulation frequency. This sampled version of $M(\tau)$ serves the role of $g_1(\tau)$ of the standard DWS experiment, giving us the decay of correlation introduced by Brownian particles within the ROI.

Before we move to the section describing the experiments, in the following we introduce our second problem of detecting another dynamic inhomogeneity, flow through a capillary buried within a turbid medium.

2.3.2 Detection and estimation of liquid flow through a capillary hidden in a turbid medium

We would like to interrogate a turbid, flesh-like object with a coherent light beam to detect, and if possible measure, fluid flow through a hidden capillary tube. It is easy to guess the motivation which is owing to its application in biomedical imaging. In as much as the flow presents itself as a localized dynamic inhomogeneity, detection of it is quite in line with the thrust of main issue discussed in this chapter. Since the decay of the detected speckle modulation, $M(\tau)$, depends solely on the dynamics within the ultrasound focal volume used to scan the object, earlier designated the ROI, this decay, and the evaluated growth of MSD with time will help one discern whether the ROI intercepts the flow or not. The MSD follows a τ^α growth with $\alpha = 1$ in the case of pure Brownian motion of particles in the ROI, and ideally $\alpha = 2$ in the case the particles co-flow with the liquid in the capillary, identification of the presence of flow can be made from the measured value of α. (Usually, because of mixing of Brownian motion with the movement associated with the flow, $1 < \alpha < 2$.) Since assessment of fluid flow has been pursued with the standard DWS with MSD evaluated from a global measurement of $g_1(\tau)$, the discriminability based on α through measuring modulation depth is shown to be far superior simply because of the confinement of measurement to a smaller region, the ROI.

2.3.2.1 Theoretical considerations

The model that is used to study correlation propagation through a turbid medium is the CDE (correlation diffusion equation) which connects $G(\mathbf{r}, \tau)$ to the optical properties and dynamics of the object, which is

$$-\nabla \cdot \kappa \nabla G(\mathbf{r}, \tau) + (\mu_a + 1/3\mu_s' k_0^2[\langle \Delta r^2(\mathbf{r}, \tau)\rangle]_b)G(\mathbf{r}, \tau) = S_0(\mathbf{r}_0) \qquad (2.14)$$

Here, almost all the parameters are defined earlier in chapter 1. We would like to reiterate that $1/3\mu_s' k_0^2[\langle \Delta r^2(\mathbf{r}, \tau)\rangle]_b$ encapsulates the background dynamics of the scattering centres through their average MSD and $S_0(\mathbf{r}_0)$ is the strength of the isotropic point source at $\mathbf{r}_0$. Where there is no embedding of the capillary the dynamics is owing to Brownian motion; inside the capillary it is due to both Brownian motion as well as the liquid flow. To arrive at this dynamics, capturing also the non-local, history-dependent behaviour of a representative system particle, a GLE has been used with a multiplicative noise term in the restoration term accounting for such indeterminacy (see section 2.2.2). Solution of the GLE is one of the routes to reach the MSD, as discussed in section 2.2.3. There is an alternative though, through solving equation (2.14) for $G(\mathbf{r}, \tau)$ and connecting it to $\langle \Delta r^2(\mathbf{r}, \tau)\rangle$ through

$$G(\mathbf{r}, \tau)|_{\mathbf{r}\in\partial\Omega} = \exp\left[-\left(\frac{L}{l^*}\right)^2 k_0^2\langle \Delta r^2(\mathbf{r}, \tau)\rangle\right] \tag{2.15}$$

Here, L is the average thickness of the object and $\langle \Delta r^2(\mathbf{r}, \tau)\rangle$ is averaged over all $\mathbf{r}$ intercepted by photon diffusion paths through the object. On the other hand, if the decay of $M(\mathbf{r}, \tau)|_{\mathbf{r}\in\partial\Omega}$ is measured, we have

$$M(\mathbf{r}, \tau)|_{\mathbf{r}\in\partial\Omega} = \exp\left[-\left(\frac{L_{\mathrm{ROI}}}{l^*}\right)^2 k_0^2\langle \Delta r^2(\mathbf{r}, \tau)\rangle\right] \tag{2.16}$$

where L_{ROI} is the average thickness of the ROI intercepted by photon paths and $\langle \Delta r^2(\mathbf{r}, \tau)\rangle$ is averaged over all diffusion paths intercepting the ROI. Depending on the measurement, we invert either equations (2.15) or (2.16) for $\langle \Delta r^2(\mathbf{r}, \tau)\rangle$, one is the average for the entire object and the other more local, average over just the ROI.

However, in order to have equation (2.16) for $M(\mathbf{r}, \tau)$ there should be ultrasound insonification which produces a perturbation in the ROI, the focal volume of the ultrasound transducer. This additional dynamics perturbs $G(\mathbf{r}, \tau)$ to $G(\mathbf{r}, \tau) + G^\delta(\mathbf{r}, \tau)$ and equation (2.14) modifies to

$$\begin{aligned}\nabla \cdot \kappa\nabla(G + G^\delta)(\mathbf{r}, \tau) + (\mu_a + (1/3)\mu_s'\mu_s' k_0^2\{[\langle \Delta r^2(\mathbf{r}, \tau)\rangle]_b + \chi_I[\langle \Delta r^2(\mathbf{r}, \tau)\rangle]_{\mathrm{in}}\} \\ + \chi_I p(\mathbf{r})\sin^2\left(\frac{\omega_f\tau}{2}\right)(G + G^\delta)(\mathbf{r}, \tau) = S_0(\mathbf{r}_0)\end{aligned} \tag{2.17}$$

Here, $p(\mathbf{r})$ is the amplitude of oscillation of the scattering particles through ultrasound forcing, averaged over $(l^*)^3$ and $G^\delta(\mathbf{r}, \tau)$, in essence, represents the sinusoidal modulation on the decaying $G(\mathbf{r}, \tau)$. Also, χ_I is the characteristic function of I the insonified ROI, where the unperturbed dynamics is represented by $[<\Delta r^2(\mathbf{r}, \tau)>]_b + [<\Delta r^2(\mathbf{r}, \tau)>]_{\mathrm{in}}$. (We note in passing that $\sin^2\left(\frac{\omega_f\tau}{2}\right)$ is an inadequate representation of the ultrasound-induced dynamics. For a more accurate representation, one should take into account the kinematical and material aspects of embedded scattering particles, and their interaction, in a visco-elastic medium.) Equations (2.14) and (2.17) come with Robin boundary conditions which for equation (2.17) is

$$(G + G^\delta)(\mathbf{r}, \tau) + \kappa\frac{\partial(G + G^\delta)(\mathbf{r}, \tau)}{\partial\hat{n}} = 0 \tag{2.18a}$$

We also have boundary conditions to be met on the surface S_C of the capillary tube:

$$G^{\text{in}}(\mathbf{r}, \tau) = G^{\text{out}}(\mathbf{r}, \tau) \text{ and } \hat{\text{n}} \cdot \nabla G^{\text{in}}(\mathbf{r}, \tau) = \hat{\text{n}} \cdot \nabla G^{\text{out}}(\mathbf{r}, \tau) \tag{2.18b}$$

Here, $G^{\text{in}}(\mathbf{r}, \tau)$ and $G^{\text{out}}(\mathbf{r}, \tau)$ are, respectively, $G(\mathbf{r}, \tau)$ inside and outside of the tube, and $\hat{n}$ is the outward normal on S_C. With these and the material parameters as input we can solve equation (2.17) for $G(\mathbf{r}, \tau) + G^{\delta}(\mathbf{r}, \tau)$, which is the 'forward solution' to the correlation diffusion problem presented in the mentioned equation. From $G(\mathbf{r}, \tau) + G^{\delta}(\mathbf{r}, \tau)$ one can arrive at $M(\mathbf{r}, \tau)$ by taking the Fourier transform magnitude at ω_f. As discussed in section 2.3.1 $M(\mathbf{r}, \tau)$ is sampled in τ through the rectangular window which was slid across $G(\mathbf{r}, \tau) + G^{\delta}(\mathbf{r}, \tau)$ for taking evolutionary power spectrum through windowed Fourier transform.

That $M(\mathbf{r}, \tau)$ is indeed the correlation decay suffered by local Brownian particles within the ROI can be verified by combining equations (2.14) and (2.17) and rewriting it as a perturbation equation. Subtracting equation (2.14) from equation (2.17) and rearranging terms we get an equation in $G^{\delta}(\mathbf{r}, \tau)$, which is

$$\begin{aligned}&-\nabla \cdot \kappa \nabla G^{\delta}(\mathbf{r}, \tau) + (\mu_a + (1/3)k_0^2\mu_s'[\langle \Delta r^2(\mathbf{r}, \tau)\rangle]_b)G^{\delta}(\mathbf{r}, \tau) = \\ &-\left(\left(\sin^2\left(\frac{\omega_f \tau}{2}\right)p(\mathbf{r}, \tau) + (1/3)k_0^2[\langle \Delta r^2(\mathbf{r}, \tau)\rangle]_{\text{in}}\right)\chi_I \mu_s'\right)G(\mathbf{r}, \tau)\end{aligned} \tag{2.19}$$

The boundary condition is

$$G^{\delta}(\mathbf{r}, \tau) + \kappa\frac{\partial G^{\delta}(\mathbf{r}, \tau)}{\partial n} = 0, \ \mathbf{r} \in \partial\Omega \tag{2.20}$$

We point out that the above approximation is valid only if the acoustic force is small enough for first Born approximation to hold. Therefore, the external force has to be kept small and thereby claim the advantage that the flow is undisturbed by the external perturbation. How small is debatable, but should be only a small fraction of the force the flow exerts, which can be arrived at from the conditions for Born approximation to hold. Examination of equation (2.19) suggests the source driving the propagation is a distributed one, weighted by $G(\mathbf{r}, \tau)$ times a sinusoidal modulation coupled with MSD, both confined spatially to the ROI. This sinusoidal modulation will be reflected on the dependent term, $G^{\delta}(\mathbf{r}, \tau)$ which decays owing to the presence of $[\langle \Delta r^2(\mathbf{r}, \tau)\rangle]_{\text{in}}$ (stochastic term); since the source is confined to the ROI, the resulting decay can be used to infer the local MSD. Information extracted from the MSD can be used to ascertain whether the capillary is present or not in the current position of the ROI. We take this route to numerically compute the decay of $M(\mathbf{r}, \tau)$ and through it the growth of $\langle \Delta r^2(\mathbf{r}, \tau)\rangle$. If the growth of $\langle \Delta r^2(\mathbf{r}, \tau)\rangle$ with τ approaches τ^2 then the ROI must be intercepting the capillary where, as per design in the experiments to follow, there is a Poiseuille flow. On the other hand, if it approaches τ then ROI is in the background medium with Brownian dynamics. Analyzing the experimental data in section 2.4.2, we see that α in τ^{α} never reaches 2 even when the ROI is inside the capillary. This is owing to the mixing of Brownian dynamics with the flow and the ever present anomalous diffusion in Poiseuille flow. However, as we prove with experiments, the ultrasound-assisted technique provides better discriminability of the presence of the inhomogeneity than the one without using tagging by ultrasound.

2.4 Experiments

2.4.1 Recovery of visco-elastic spectra from an inhomogeneous tissue-like object

Here we describe the first experiment demonstrating the application of localized ultrasound insonification for detecting and measuring dynamic inhomogeneities: DWS-based micro-rheology in an inhomogeneous medium confined to where the ultrasound forcing is localized. Our objective here is first to prove that the decay of $M(\tau)$ is caused indeed by the Brownian particles within the ROI and then apply the modified Stokes–Einstein relation [3, 4] to compute the visco-elastic spectra of the material of the ROI. The object in our experiments is a composite phantom which is the sandwich of two slabs, each made of (poly) vinyl alcohol (PVA) differing in visco-elastic properties. (PVA mimics, not completely but certain aspects of, healthy as well as diseased human breast tissue.) For optical scattering coefficient the match is far from ideal, with the normal breast tissue presenting a μ_s in the range of 10–50 mm^{-1} in the near-infrared wavelength range and for the PVA samples it lies within the ceiling of 25 mm^{-1}. But for shear modulus the match can be made perfect, both for healthy as well as cancerous samples, by adjusting certain parameters whilst fabricating the PVA [19]. The dimensions of the two slabs in the sandwich are 30 mm × 50 mm × 8 mm each, one of shear modulus 11 kPa and the other 23 kPa. The PVA gel, since the scattering centres are bound to each other, does not exhibit ergodic behaviour; however, by sandwiching to a cuvette containing 'free' polystyrene spheres in glycerol (in our experiments, 4% solution of spheres of diameter ~2 μm; $l^*\simeq 737$ μm, as per Mie theory) to the composite polymer slab can render the entire object to exhibit ergodic behaviour [20].

Fabrication of the PVA slabs for bio-imaging work followed the recipe given in [19]. A quick summary of the results achieved is quoted below; for details the reader is referred to [19] and the references therein. By varying what is referred to as 'freeze–thaw cycles' one could get a range of shear moduli for the slab, roughly from 11 kPa to 97 kPa. With shear modulus increase μ_s also increases, from approximately 10 to 25 mm^{-1} at $\lambda = 633$ μm. If further increase in μ_s is desired, then external sources of scattering, polystyrene beads or TiO_2 powder, need to be added in the stock solution. For fixing μ_a an appropriate quantity of India ink is also added.

2.4.1.1 The experimental setup, data gathering and processing

The aim of the experiment is to record $g_1(\mathbf{r}, \tau)$ at a detector location gathering photons from the boundary of the slab, modulated also by acoustic pressure from ultrasound transducers. The ROI, in the present case, consists of the intersection volume of the focal regions of two synchronously driven ultrasound transducers. The transducers are driven by a dual-mode function generator driven at frequencies of 1.00 MHz and 1.00 MHz + Δf Hz where Δf, the difference frequency can be set at any value ranging from tens of Hz to a few kHz. The sinusoidal pressure signals from the two transducers 'mix' at the common focal volume producing a low-frequency acoustic pressure wave driving the scattering centres in the ROI. The advantage of this arrangement is that we have increased penetration of acoustic waves at higher frequencies, whilst allowing low-frequency acoustic radiation force at the ROI which could be deep within the object. The low-frequency forcing excites the natural frequencies of the ROI which for the material of the slab is characteristically of low frequency.

The experimental setup is shown in figure 2.1. The composite object together with the cuvette is illuminated by an unexpanded beam of light from He–Ne laser, emitting 15 mW of light at 633 nm (from Thorlabs, HRR170). Focused dual-beam ultrasound radiation is from a confocal transducer, part energized at 1.00 MHz and the other part at 1.00 MHz + Δf Hz, where Δf is selected to be 1 kHz. The transducer is aligned with the help of an x–y–z translation stage so that the common focal volume (approximately, elongated hyperboloidal shape) intercepts the object at the required region perpendicular to the direction of illumination. The composite object has only two regions: one softer with shear modulus of 11 kPa and the other stiffer at 23 kPa. A water bath submerges both the object and the ultrasound transducer for proper impedance matching.

From the speckle pattern available at the light exit plane a single speckle is captured in a single-mode fibre with the alignment achieved through a motorised translation stage. We have maintained the ultrasound forcing small; typical value of the power delivered from each section of the transducer is 15 mW. With this, the Peclet number is maintained low and temperature rise in the ROI through

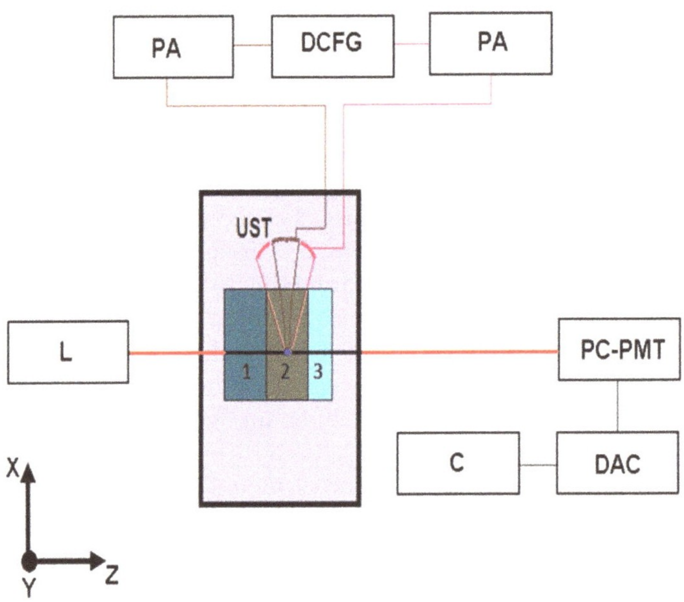

Figure 2.1. Light emitted from laser source L, illuminates the ROI insonified by confocal ultrasound transducer (UST). The scattered intensity of light is detected by the detector (PC-PMT) and is given to the correlator DAC and then to a computer C. The UST is driven by power amplifier (PA) that takes input from a dual-channel function generator (DCFG). The sample consists of two slices of PVA, of different storage modulus values, 11 kPa (1) and 23 kPa (2), and an ergodic medium (3). Reprinted figure with permission from [8] © (2014) by the American Physical Society.

ultrasound is negligible. With the low Peclet number, the signal-to-noise ratio (SNR) of the desired sinusoidal modulation in the data is small; therefore, it was essential to maximize the SNR by careful alignment of the fibre to the speckle maximum. The captured light is detected using a photon-counting PMT (Hamamatsu HR-7360-03). The output from the PMT is given to a hardwired autocorrelator (Flex 021d from DAC, www.correlator.com). In the experiment we gather intensity variation with respect to time, uniformly sampled (i.e. using photon-mode of data gathering using the autocorrelator, as also discussed in section 2.3.1.2). The acoustic force at the beat frequency of 1 kHz provides an adequate 'carrier frequency' so that visco-elastic spectra at the frequency bandwidth of interest here can be extracted from the data. The frequency is also low enough to ensure shear-dominated vibration at the ROI.

The details of the experiment done are given below: We gather a number of samples of evolution of intensity with time corresponding first to the ergodic medium in the cuvette and then the sandwich of the composite slab and the ergodic medium. For the composite object two sets of data are gathered: (1) corresponding to the ultrasound forcing in the 23 kPa portion and (2) the other when the forcing is in the 11 kPa portion. From the gathered intensities normalized intensity autocorrelation, $g_2(\tau)$, corresponding to the three cases under consideration are computed. From $g_2(\tau)$ the modulus of the amplitude autocorrelation $|g_1(\tau)|$ is found using Siegert relation $g_2(\tau) = 1 + f\,|g_1(\tau)|^2$ wherein the coupling constant f can be computed from $g_2(0)$. Once $g_1(\tau)$ for the phantom-cuvette double cell, and the ergodic medium in the cuvette alone are available, the multiplication rule set forth in [20] can be employed to retrieve $g_1(\tau)$ for the composite phantom. Two samples of the retrieved $g_1(\tau)$ are shown in figure 2.2, with the intersecting focal volumes in the two regions making the composite slab.

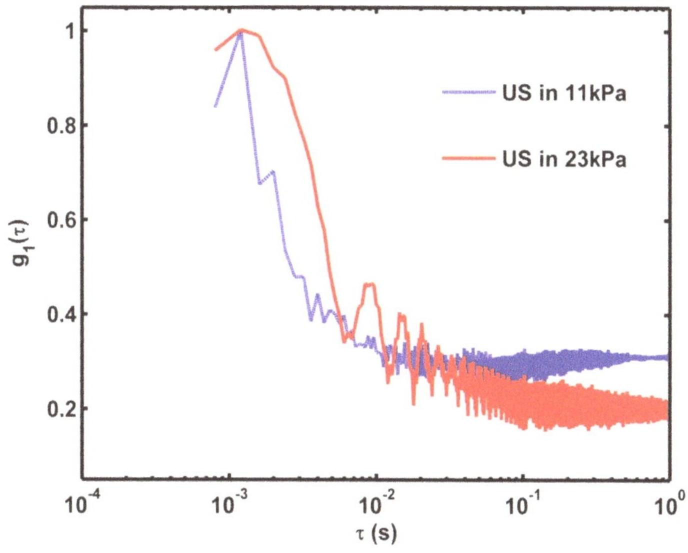

Figure 2.2. Normalized filed autocorrelation plots for individual phantoms, with US in 11 kPa (blue) and 23 kPa (red), extracted using the multiplication rule. Reprinted figure with permission from [8] © (2014) by the American Physical Society.

We now proceed to apply short-time Fourier transform on the obtained $g_1(\tau)'s$ in order to extract the variation of $M(\tau)$ with τ. A rectangular window of length 4 ms is slid over $g_1(\tau)$ with a time step of 400 μs and at each position, τ_i, Fourier transform magnitude at 1 kHz is obtained, which is $M(\tau_i)$. (Since the modulation at beat frequency (1 kHz) acts like a carrier from which visco-elastic spectra are extracted, 500 Hz sets a limit for the frequency of the signal extracted. To satisfy sampling theorem requirements with bandwidth to spare we kept the sampling time to be 400 μs.) The plots of $M(\tau)$ vs τ for the two samples are in figure 2.3. It is seen that for small values of τ the SNR is poor and the data unreliable. However, for τ from 0.1 s till 1.0 s the SNR is good enough to show forth the dependence of the decay on the location of the ROI.

Equation (2.16) establishes the connection between $M(\tau)$ and $\langle\Delta r^2(\mathbf{r}, \tau)\rangle$ pertaining to the Brownian particles within the ROI. This equation is inverted to arrive at the evolution of $\langle\Delta r^2(\mathbf{r}, \tau)\rangle$ with τ. These are shown in figures 2.4 and 2.5, where for comparison $\langle\Delta r^2(\mathbf{r}, \tau)\rangle$ vs τ graphs obtained from $g_1(\tau)$, i.e. without using ultrasound, are also shown. The generalized Stokes–Einstein relation (GSER) relates $\langle\Delta r^2(\mathbf{r}, \tau)\rangle$ to $G^*(\omega)$ the frequency-dependent complex modulus of elasticity [3, 4]:

$$G^*(\omega) = \frac{k_B T}{\pi \xi i \omega F\{<\Delta r^2(\mathbf{r}, \tau)>\}} \tag{2.21}$$

Here, T stands for temperature in Kelvin, F for Fourier transform and k_B Boltzmann's constant. Moreover, τ is the average diameter of the scattering centres. From this, one can easily compute the storage- and loss-moduli spectra using [8],

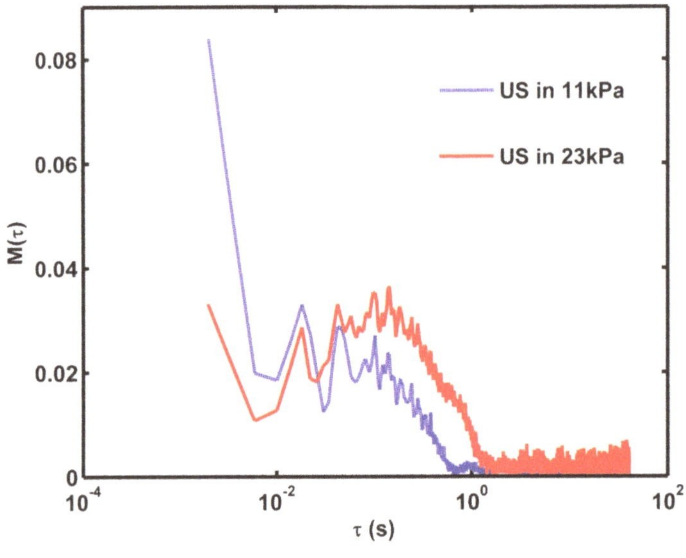

Figure 2.3. Modulation depth decay when the ROI is in either of the two regions in the composite phantom. Reprinted figure with permission from [8] © (2014) by the American Physical Society.

$$G'(\omega) = G(\omega)\{1/[1 + \beta']\}\cos\left[\frac{\pi\alpha'(\omega)}{2} - \beta'(\omega)\alpha'(\omega)\left(\frac{\pi}{2} - 1\right)\right] \tag{2.22}$$

$$G''(\omega) = G(\omega)\{1/[1 + \beta']\}\sin\left[\frac{\pi\alpha'(\omega)}{2} - \beta'(\omega)(1 - \alpha'(\omega))\left(\frac{\pi}{2} - 1\right)\right] \tag{2.23}$$

Here,

$$G(\omega) = \frac{k_B T}{\pi\varsigma\left\langle \Delta\mathbf{r}^2\left(\frac{1}{\omega}\right)\right\rangle\Gamma[1 + \alpha(\omega)][1 + \beta(\omega)/2]} \tag{2.24}$$

which is also called the frequency dependent elastic modulus. Also, $\alpha(\omega)$ and $\beta(\omega)$ are the first- and second-order logarithmic time derivatives of $\langle\Delta r^2(\mathbf{r}, \tau)\rangle$ and $\left\langle \Delta\mathbf{r}^2\left(\frac{1}{\omega}\right)\right\rangle$ is the MSD evaluated at $\tau = \frac{1}{\omega}$. Moreover, $\Gamma(\cdot)$ is the gamma function, $\alpha'(\omega)$ and $\beta'(\omega)$ are the first- and second-order logarithmic derivatives of $G(\omega)$ and ς is the radius of a typical Brownian particle. A second-order polynomial fit using a sliding Gaussian window is used to smooth the MSD data and to obtain $\alpha(\omega)$ and $\beta(\omega)$. The values used are 6.6 and 5 for the storage modulus values of 11 and 23 kPa, respectively. In the next section we discuss these results of retrieval of $\langle\Delta r^2(\mathbf{r}, \tau)\rangle$, $G'(\omega)$ and $G''(\omega)$ from $M(\tau)$ and compare them with those obtained from $g_1(\mathbf{r}, \tau)$ from homogeneous slabs of the same mechanical properties as the parts of the composite phantom. In addition, $\langle\Delta r^2(\mathbf{r}, \tau)\rangle$ is computed through simulation, solving equation (2.4) as discussed in section 2.2.3.

2.4.1.2 Results and discussion

Figures 2.4 and 2.5 give comparisons of evolution of MSD with τ obtained through simulation and experiments. These results substantiate our claim that the decay in amplitude correlation suffered by light owing to fractional Brownian motion within the ROI is indeed captured by the decay in $M(\mathbf{r}, \tau)$. This is verified by the match of MSDs, those obtained from the measured $M(\mathbf{r}, \tau)$ with those from simulations as well as from $g_1(\mathbf{r}, \tau)$ measured from a homogeneous slab. Simulation results, it is observed, match better with experimental results from a homogenous phantom done without ultrasound forcing than those form ultra-sound-assisted local measurements. This is quite evident when the plateau is reached for relatively large time delays. Possible reasons for this behaviour are: (1) acoustic absorption in the vicinity of the ROI, even though small, can upset the thermal equilibrium, which pushes the ROI to a higher equilibrium temperature. This means the MSD will settle at a higher plateau; (2) mixing of stochastic dynamics with the deterministic, acoustically-induced dynamics, which influences the large-time behaviour of the MSD. By taking care to see that the acoustic force is small, heat generation and the consequent rise in temperature can be avoided. A model to account for this mixing is the GLE developed in section 2.2.2 with a

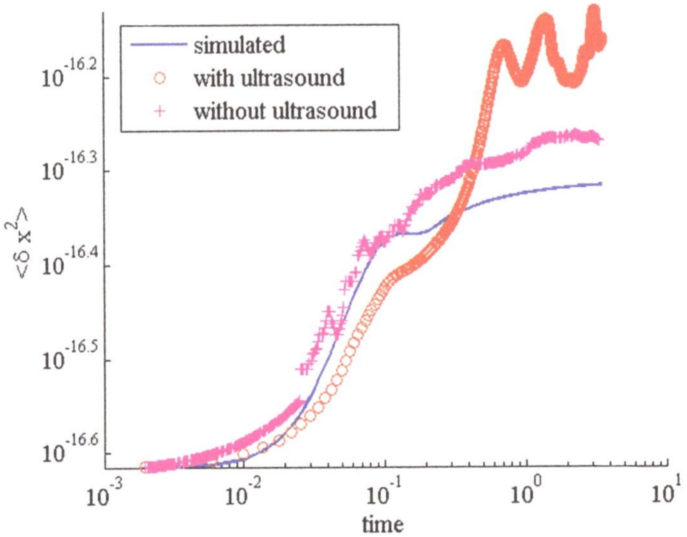

Figure 2.4. The MSD versus τ graphs, simulated using equation (2.6) compared with those from experimentally measured autocorrelation with ultrasound modulation (circles) and without (+ symbols), for region 2 in the composite phantom with shear modulus of 23 kPa. Reprinted figure with permission from [8] © (2014) by the American Physical Society.

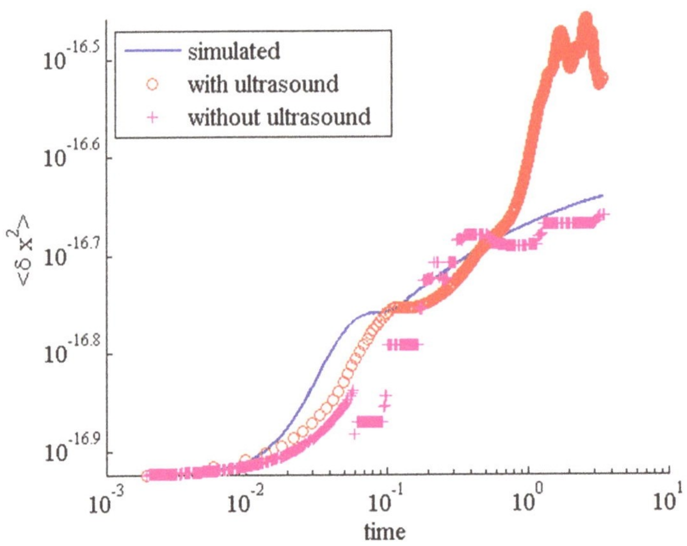

Figure 2.5. Same as figure 2.4 done for region1 of the composite phantom with shear modulus of 11 kPa. Reprinted figure with permission from [8] © (2014) by the American Physical Society.

multiplicative noise term in the restoration term. In section 2.2.3 we touched upon solving this equation to arrive at how MSD evolves with time, using a change of measure. Through this, it is shown that (see figures 2.4 and 2.5) contributions to the growth of MSD from external forcing assume significance only for large values of τ.

Consequently at the early transients, evolution of MSD matches perfectly in the two cases of with and without insonification.

In computing the visco-elastic spectra one uses only the early part of the MSD growth, and hence the computed $G'(\omega)$ and $G''(\omega)$ in the two cases also agree quite well. The arrived at visco-elastic spectra are shown in figures 2.6 and 2.7, including those obtained from individual homogeneous samples, which are obtained for the sake of comparison. The match is quite good in the case of $G'(\omega)$, but not quite so in the case of $G''(\omega)$. There is an inherent uncertainty in the DWS-based measurement of the viscous part of the modulus spectrum (either with or without the ultrasound) which would answer this lack of fidelity. Since 1 kHz is the sampling frequency we expect reliable measurements only up to 500 Hz.

2.4.2 Detection and estimation of liquid flow through capillary

The second experiment is to demonstrate how a dynamic inhomogeneity in the shape of flow in a capillary tube can be detected relatively easily using modulation decay measurement from an ultrasound-assisted measurement. The experimental setup is shown in figure 2.8. The object under study is a PVA phantom of dimensions $60 \times 90 \times 30\ \text{mm}^3$ with a 10 mm diameter quartz tube embedded at a depth of 15 mm and running through its entire breadth horizontally. The scattering coefficients, μ_s and μ_s', are measured and found to be $7.2\ \text{mm}^{-1}$ and $0.8\ \text{mm}^{-1}$, respectively. The absorption coefficient is left small at its native value of

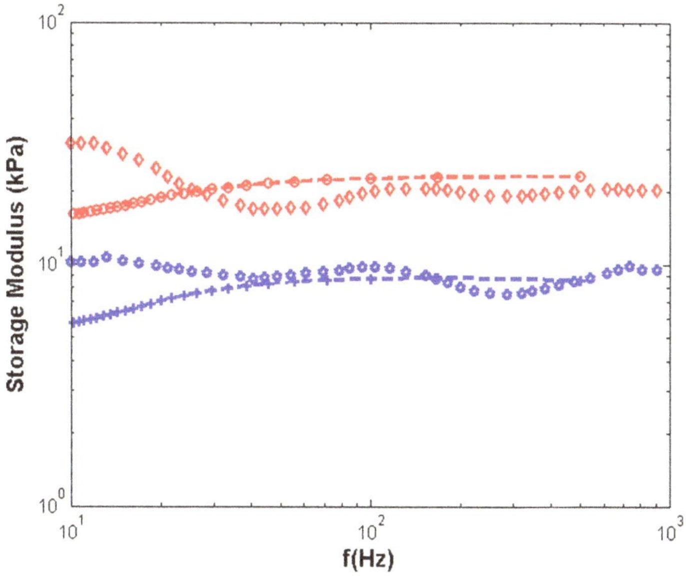

Figure 2.6. Comparison of storage moduli spectra obtained by localized ultrasound assisted DWS from the measured $M(\tau)$ decay (-+-+- when the US forcing is in the 11 kPa block and -o-o- when the US forcing is in the 23 kPa block) with those obtained from the usual DWS-based measurement of $g_2(\tau)$ decay for individual homogeneous phantoms (* * for 11 kPa block and ◊ ◊ for the 23 kPa block). Reprinted figure with permission from [8] © (2014) by the American Physical Society.

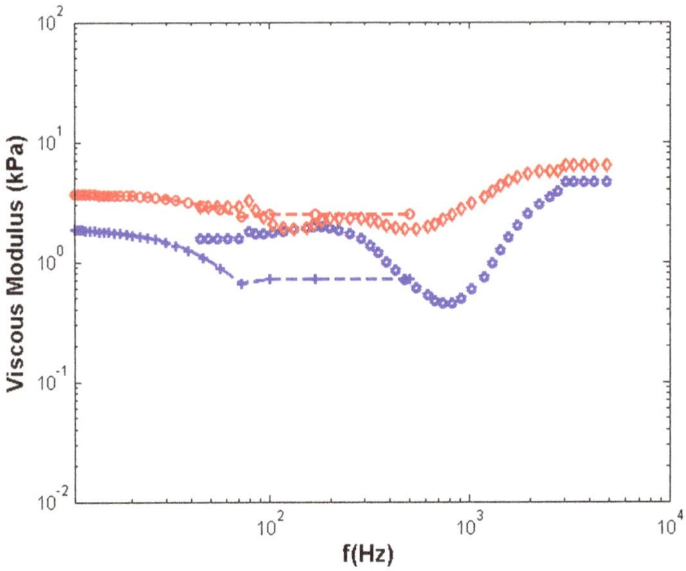

Figure 2.7. Comparison of loss moduli spectra obtained by localized ultrasound assisted DWS from the measured $M(\tau)$ decay (-+-+- when the US forcing is in the 11 kPa block and -o-o- when the US forcing is in the 23 kPa block) with those obtained from the usual DWS measurement of $g_2(\tau)$ decay with homogeneous phantoms (* * for 11 kPa block and ◊ ◊ for the 23 kPa block). Reprinted figure with permission from [8] © (2014) by the American Physical Society.

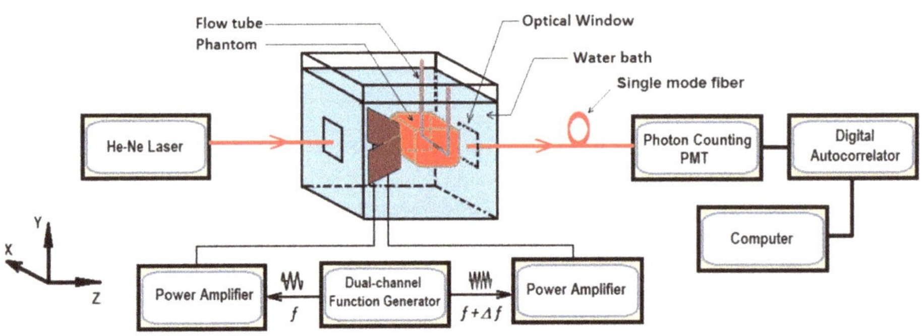

Figure 2.8. The experimental setup. The *PVA* phantom (a slab of dimensions 60 × 90 × 30 mm^3), containing the flow pipe at a depth of 15 mm from the surface of the slab which receives the light, i.e. the x–y plane in the figure, is immersed in a water bath for acoustic impedance matching. Two focusing US transducers with provision for fine mechanical alignment (not shown) are driven synchronously by a dual-channel function generator at frequencies f and $f + \Delta f$ whose focal regions intersect at their waist regions at an angle of 60°. The laser is input along the line bisecting this angle directed toward the intersecting volume. A single speckle formed at the exit is captured by a single mode fibre and given to a photon counting PMT which is connected to a digital autocorrelator and a computer. Reprinted figure with permission from [8] © (2014) by the American Physical Society.

0.25×10^{-3} mm^{-1}. The background medium is assumed free of any other movement than temperature-induced Brownian motion; therefore, when the ROI is entirely confined to the background, the term $(1/3)k_0^2\mu_s'[\langle\Delta r^2(\mathbf{r}, \tau)\rangle]_b$ in equation (2.19) can be written as $2\mu_s k_0^2 D_B^{out}(\frac{\tau}{l^*}) \equiv 2\left(\frac{\tau}{\tau_0^{out} l^*}\right)$ where $\tau_0^{out} = (k_0^2 D_B^{out})^{-1}$ and D_B^{out} is the particle diffusion coefficient in the PVA slab. On the other hand when the probing ROI is within the capillary tube, the dynamics has contributions from Poiseuille flow within as well as the ever present Brownian motion, characterized by particle diffusion coefficient D_B^{in} in the liquid. In this case, $(1/3)k_0^2\mu_s'[\langle\Delta r^2(\mathbf{r}, \tau)\rangle]_{\text{in}}$ becomes $2\left(\frac{\tau}{\tau_0^{\text{in}} l^*}\right) + \frac{2}{l^*}\left(\frac{\tau}{\tau_f}\right)^2$. Here, $\tau_0^{\text{in}} = (k_0^2 D_B^{\text{in}})^{-1}$, $\tau_f = \frac{\sqrt{30}}{k_0 l^* \Gamma_1}$ and $\Gamma_1 = \frac{32Q}{\pi d^3}\left(\frac{1}{3} - \frac{2}{\pi^2}\right)$. Moreover, d and Q are the diameter of the capillary and volume flow rate of fluid through it, respectively [21–23].

2.4.2.1 The experimental setup

The setup in figure 2.8, is almost identical to the one shown earlier in figure 2.1, except that there is provision for flow through the embedded capillary in the object. The insonification is from two separate ultrasound transducers which are mounted on translation stages for precise alignment of the location of the region of intersection of the focal volumes (also called the ROI) within the object. The storage modulus of the PVA slab is kept at 11 kPa, for which the average particle diffusion coefficient (D_B^{out}) is experimentally measured to be 1.5×10^{-10}mm^{2}s^{-1} at 27 °C. Turbid water is circulated through the embedded quartz tube with a flow pump which is adjusted for a desired volume flow rate. Turbidity in water is created by adding polystyrene balls of average diameter 2.0 μm at a volume fraction of 0.004 which gave an l^* of approximately 1.25 mm as per Mie theory [24]. For the turbid liquid D_B^{in} is computed to be 2.46×10^{-7} mm^{2}s^{-1} at 27 °C. With the help of pump and valves the volume flow-rate can be adjusted as per requirements. The rate is approximately ascertained by measuring the volume discharged in a fixed time. The object with the embedding is immersed in water bath for acoustic impedance matching during insonification (figure 2.8).

There are two ultrasound transducers, identical in all aspects, focusing acoustic waves to a common intercept region within the object. Both are of focal length 50 mm and *f/no* 1.72, and operated in continuous mode (Make: Panametrics V394-SU-F50MM-PTF). The transducers are aligned in such a way that their axes intersect at an angle 60° at the thinnest region in its focal volume which is approximately hyperboloidal in shape. The transducers are driven synchronously by a dual-channel function generator, one at 1.0 MHz and the other at 1.0 MHz + Δf Hz where Δf can be tuned in the range 100 Hz – 400 kHz.The result at the intersection, the ROI, is a low-frequency acoustic forcing at the beat frequency Δf Hz. By proper alignment, the ROI can be made small enough to lie fully within the capillary tube. As discussed in chapter 1, smallness of the ROI pushes up the SNR of the speckle modulation, which renders the detection of the measurement signal easier. The externally introduced acoustic force is kept low so that the Peclet number is small. In spite of this, even a small acoustic force in the fluid upsets the dynamics of the flow and the scattering centres in the flow suffer micro-rotations. Effective reduction in l^*, and the random addition to displacement suffered by the particles consequent to micro-rotations combine to render the MSD extracted somewhat lower than its actual value, even when it is extracted from $M(\tau)$.

Consequently, the flow velocity computed from the MSD is also much lower than its actual value (see section 2.4.2.2). As discussed and verified in section 2.4.2.2, a means to overcome this inaccuracy is to align the ultrasound force to counter the flow and adjust the acoustic force to effectively cancel the flow for a null measurement. The MSD variation with time will be linear indicating that at dynamic equilibrium the drift is fully cancelled and the dynamics is wholly Brownian. From the force applied to achieve this, we have computed volume flow which matched reasonably well with an alternative measurement using a flow-meter.

The other parts of the experimental setup are identical to that shown in figure 2.1 for the first experiment. Light from a 15 mW He–Ne laser, unexpanded, illuminates the PVA slab with the capillary. Of all the diffuse photons, some intersect with the ROI carrying phase modulation picked-up from there. These interfere at the exit plane giving a speckle pattern with a sinusoidal modulation. A single-mode fibre, carefully aligned to capture just one speckle transmits the speckle modulation to the detector which is a photon-counting PMT, Hamamatsu, H 7360-03. The modulation in the speckle and its decay are the signals we are interested in, in the present experiment. Current from the PMT after proper signal conditioning is fed to a digital autocorrelator as in the first experiment.

Output from the autocorrelator is $g_2(\tau)$ from which the modulus of $g_1(\tau)$ is arrived at, as before. From $g_1(\tau)$, as described in sections 2.3.1.1 and 2.4.1.2, $M(\tau)$ is computed. From both, $\langle \Delta r^2(\mathbf{r}, \tau) \rangle$ is arrived at, and the rise of this second moment with delay time is our discriminant to tell the capillary from the background. The experiments reported here are done in transmission geometry. However, to push towards a commercially viable prototype one needs to work with diffuse reflected light. To demonstrate the proof-of-concept we have used the rather simpler transmission mode, though unviable as a diagnostic tool.

Prior to data gathering, the ultrasound transducers, mounted on $x - y - z - \theta$ stages, are aligned for the least interaction volume, which is also verified by ultrasound detectors. In the experiments, first the ROI is adjusted to be fully outside the capillary and then fully within. The smallness of the ROI ensures that it can be brought fully within the flowing turbid water. We measure time variation of intensity from a single speckle with the help of a single-mode fibre and photon counting PMT. The output from the PMT is used to compute $g_2(\tau)$, $g_1(\tau)$ and $M(\tau)$. Five sets of data are collected corresponding to five different flow rates (Q): 0 $\mu l\, s^{-1}$, 19 $\mu l\, s^{-1}$, 33 $\mu l\, s^{-1}$, 47 $\mu l\, s^{-1}$ and 139 $\mu l\, s^{-1}$. Even for $Q = 139\ \mu l\, s^{-1}$, the Reynolds number is well below the threshold for going over to turbulence [25]. The beat frequency Δf is chosen to be 400 kHz which enabled capture of decay time even for the highest flow-rate. The data gathering took typically ~1800 s, in which time approximately 3×10^6 sets of intensity data over time intervals of 600 μs are gathered and averaged. From this, the average $g_2(\tau)$ is computed. This extraordinary long data gathering time will exclude this method from exploitation in practical clinical setting. Since the intensity decay time of photons exiting the PVA slab is of the order of hundreds of seconds, the data gathering time has to be correspondingly large. One way of reducing this time is to use parallel detection as is done in UMOT to improve the

SNR. From both $M(\tau)$ and $g_1(\tau)$ we have recovered the evolution of $\langle \Delta r^2(\mathbf{r}, \tau) \rangle$ with τ. In addition, we have also computed all the above quantities by solving equations (2.17) and (2.19) assuming appropriate values of τ_0 and τ_f using values of D_B^{out} and Q. These results are shown in the plots of figures 2.9–2.11. It is observed that the change in decay owing to change in Q is more pronounced in $M(\tau)$ than in $g_1(\tau)$, which is also reflected in the ascent to the plateau of $\langle \Delta r^2(\mathbf{r}, \tau) \rangle$ versus τ plots derived from them. The results are further discussed in the next section.

2.4.2.2 Results and discussion

Figures 2.9 and 2.10 show variations of $M(\tau)$ and $g_1(\tau)$ obtained from simulations and experiments, respectively. When the capillary (with liquid flow through it) is interrogated the decay in $M(\tau)$ is much faster than the decay in $g_1(\tau)$. (This is not surprising, because $M(\tau)$ encapsulates local dynamics within the ROI, whereas $g_1(\tau)$ global; and its decay is not quite seriously affected by dynamics or changes therein unless the ROI is not buried deep within, which allows a number of photon paths intercept it.) When the flow velocity increases the decay of both increases; however, as seen in the figures, the increase in decay of $M(\tau)$ is more pronounced.

The measured $g_1(\tau)$ and $M(\tau)$ data are used in equations (2.15) and (2.16) and the corresponding rise of $\langle \Delta r^2(\tau) \rangle$ with τ are found. Here, $\langle \Delta r^2(\tau) \rangle$ is the spatial average of $\langle \Delta r^2(\mathbf{r}, \tau) \rangle$ within the ROI. These are plotted in figure 2.11. Our first observation is that the MSD's from $g_1(\tau)$'s are an order-of-magnitude smaller than those from $M(\tau)'s$. Moreover, to quantify the growth of $\langle \Delta r^2(\tau) \rangle$ with τ we had put a τ^α fit to them. The case of $\langle \Delta r^2(\tau) \rangle$ from $g_1(\tau)$ was answered by roughly $1.46 < \alpha < 1.58$ in the time interval $0 < \tau < \tau_f$ (μs) where τ_f varies from around 52 μs *to* 275 μs for

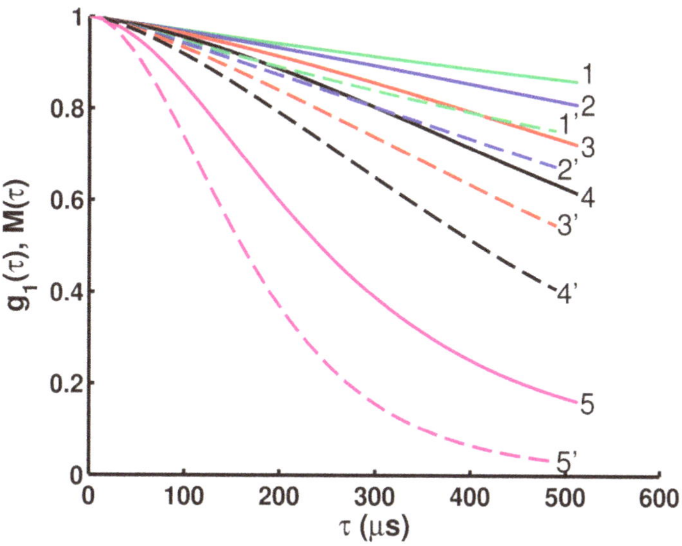

Figure 2.9. Simulated $g_1(\tau)$(-) and $M(\tau)$ (–) plots for no flow case (1,1′) are compared with that for different flow rates; 19 μl s^{-1} (2,2′), 33 μl s^{-1} (3,3′), 47 μl s^{-1} (4,4′), 139 μl s^{-1} (5,5′). Similar experimental results are shown in figure 2.10 below. Reprinted with permission from [25]. (2015) Optical Society of America.

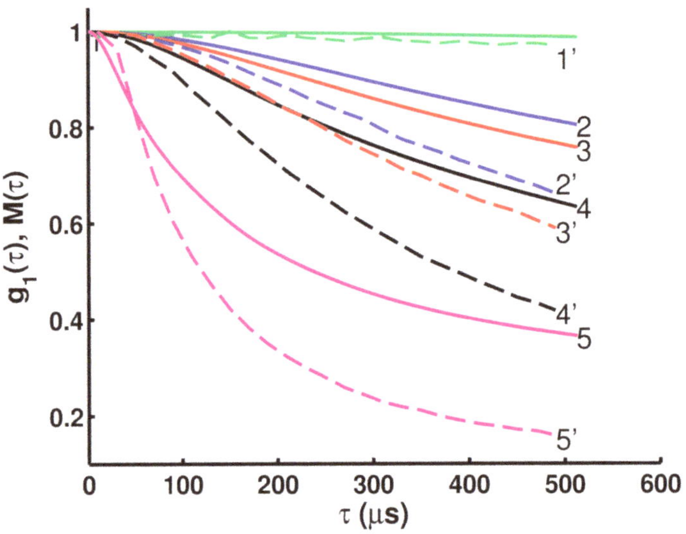

Figure 2.10. Experimental $g_1(\tau)$ (-) and $M(\tau)$ (–) plots for no flow case (1,1′) are compared with those for different flow rates; 19 μl s^{-1} (2,2′), 33 μl s^{-1} (3,3′), 47 μl s^{-1} (4,4′), 139 μl s^{-1} (5,5′). It is seen that the decay rate obtained from $M(\tau)$ is higher. Reprinted with permission from [25]. (2015) Optical Society of America.

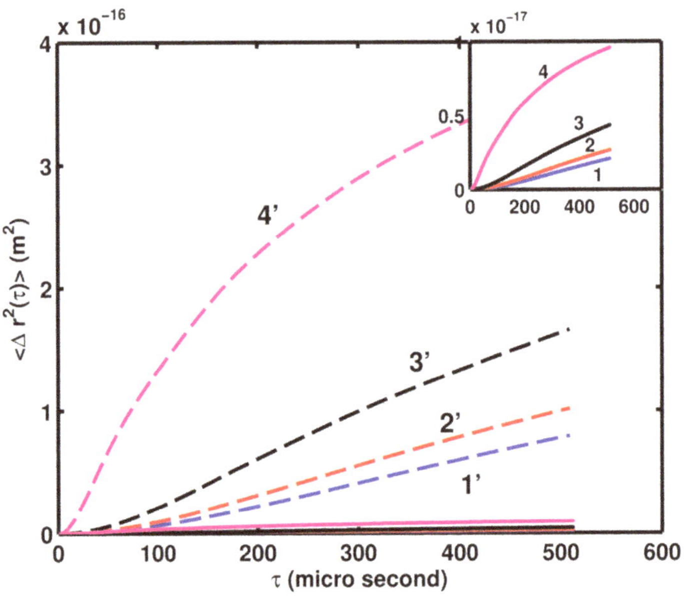

Figure 2.11. Plots of $\langle \Delta r^2(\tau) \rangle$ calculated from $g_1(\tau)$ (-) and $M(\tau)$ (-.-) for different flow rates 19 μl s^{-1} (1,1′), 33 μl s^{-1} (2,2′), 47 μl s^{-1} (3,3′), 139 μl s^{-1} (4,4′). The inset gives $\langle \Delta r^2(\tau) \rangle$ from global measurement at a different scale to bring forth the variation with τ. Reprinted with permission from [25]. (2015) Optical Society of America.

different flow-rates. Thereafter $\alpha \approx 1$ showing that for large delay times the dynamics is fully Brownian (see figures 2.12 and 2.14). In the case of MSD from $M(\tau)$ we have $1.65 < \alpha < 1.77$ for $0 < \tau < \tau_f$ with τ_f the threshold varying from 59 μs *to* 335 μs representing different flow-rates. Thereafter, as in the previous case, $\alpha \approx 1$ (see figures 2.13 and 2.15; also table 2.1).

One of the easily noticeable features of both the plots is that relatively short-pathlength photons, for which correlation exists for large time delays as well, fail to capture the signature of the flow. This could be owing to the interception of ballistic- and near-ballistic photon paths by the scattering centres orthogonal to the displacement accorded by the flow. This means, within the ROI light captures only the random dynamics, the Brownian motion, of the scattering centres. However, when the large pathlength photons capture the dynamics (in both the cases) it captures both the deterministic one from the flow and the stochastic one from Brownian motion. Also, the external force from ultrasound can trigger micro-rotation in the fluid particles which could add to the random component of the dynamics. This additional random component can hasten the transition from anisotropic to isotropic behaviour of the particles in regard to light scattering [26]. This would also bring about a reduction in l^*, which aspect is not considered in the present work. We also observe that the time-window used to compute short-time Fourier transform *en route* to computing $M(\tau)$ is shorter than the time needed for Taylor diffusion to take over which is $(D_B^{\text{in}})^{-1}$ [27]. Therefore, within the selected time-window anomalous diffusion should be accounted for, even without bringing in the mixing caused by the external forcing. What we have proven from the experimental results is that decay in $M(\tau)$ allows us to compute this mixture with a predominant contribution from the flow over

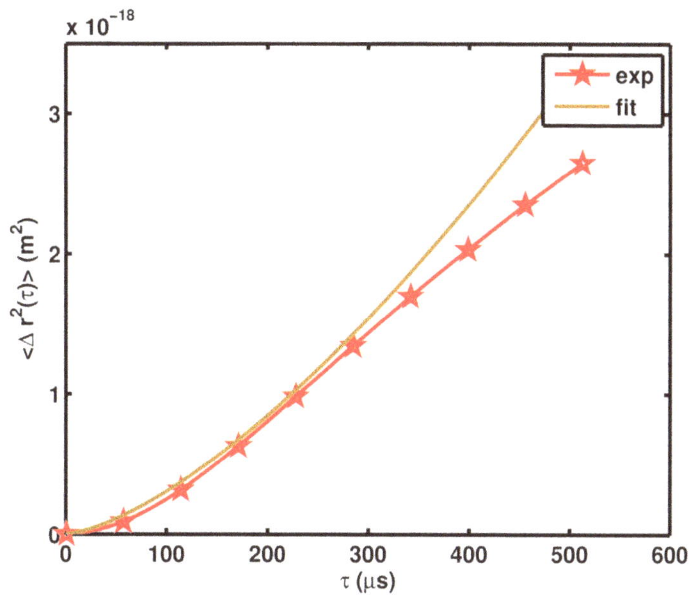

Figure 2.12. $\langle \Delta r^2(\tau) \rangle$ versus τ plot obtained from $g_1(\tau)$ for flow rate of 33 μl s^{-1}. The fit obtained using $a + b\tau^\alpha$ follows a τ^α variation with $\alpha = 1.47$ for $0 < \tau < 220$ μs. Reprinted with permission from [25]. (2015) Optical Society of America.

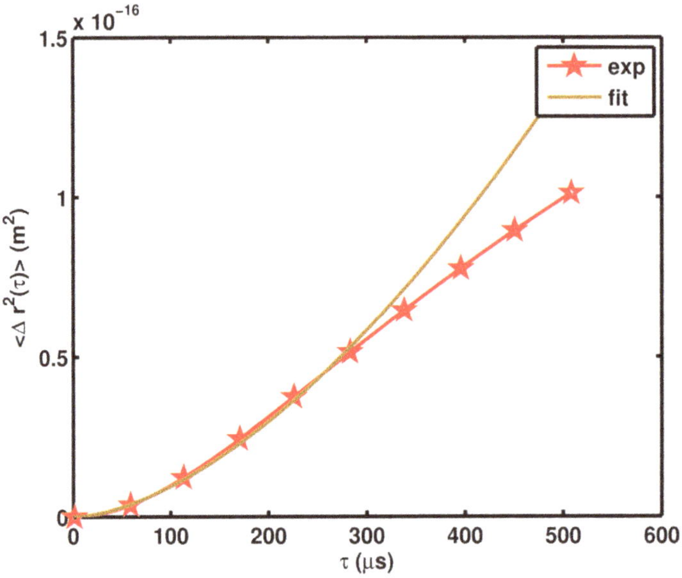

Figure 2.13. Same as figure 2.12 except that this curve is obtained from $M(\tau)$. The fit obtained using $a + b\tau^{\alpha}$ shows that for $0 < \tau < 280$ μs it follows a τ^{α} variation with $\alpha = 1.66$. Reprinted with permission from [25]. (2015) Optical Society of America.

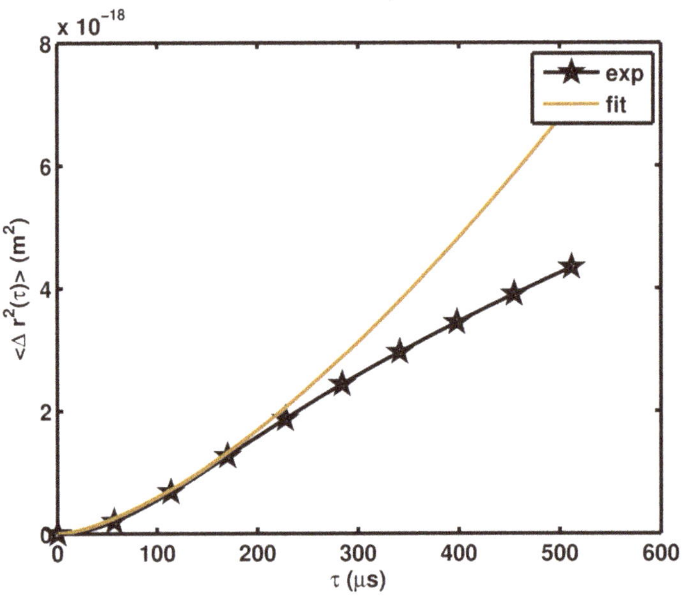

Figure 2.14. $<\Delta r^2(\tau)>$ vs τ plot obtained from $g_1(\tau)$ for flow rate of 47 μl s^{-1}. The fit obtained using $a + b\tau^{\alpha}$ follows a τ^{α} variation with $\alpha = 1.52$ for $0 < \tau < 180$ μs. Reprinted with permission from [25]. (2015) Optical Society of America.

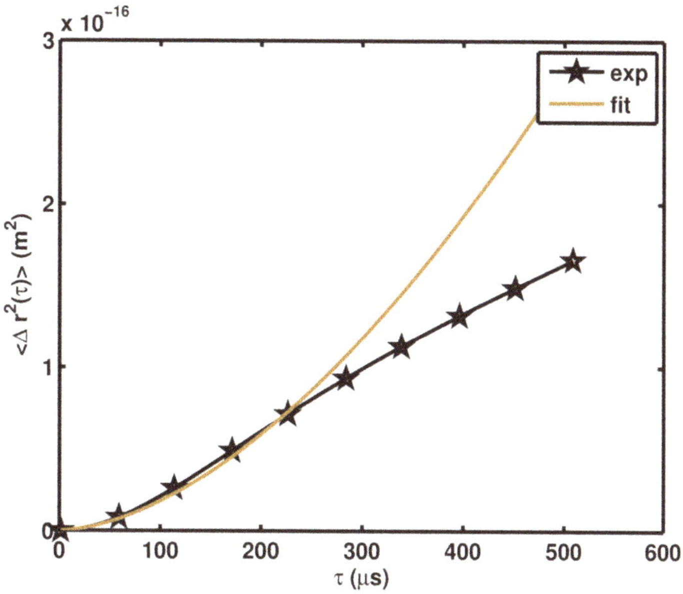

Figure 2.15. Same as figure 2.14, but extracted from $M(\tau)$. The fit obtained using $a + b\tau^{\alpha}$ shows that for $0 < \tau < 220\mu s$ it follows a τ^{α} variation with $\alpha = 1.71$. Reprinted with permission from [25]. (2015) Optical Society of America.

Table 2.1. Assuming $\langle \Delta r^2(\tau) \rangle$ follows $a + b\tau^{\alpha}$, the value of α obtained by fitting $\langle \Delta r^2(\tau) \rangle$ computed from g_1 and M for different flow rates. Reprinted with permission from [25]. (2015) Optical Society of America.

	From g_1 (τ)		From M (τ)	
Volume flow rate (μl s^{-1})	α	τ_f (in μs)	α	τ_f (in μs)
19 μl s^{-1}	1.46	275	1.65	335
33 μl s^{-1}	1.47	220	1.66	280
47 μl s^{-1}	1.52	180	1.71	220
139 μl s^{-1}	1.58	52	1.77	59

larger periods of time than is possible from $g_1(\tau)$. Since the ultrasound-assisted measurement is local this delineation of dynamics is unaffected by the presence of dynamic inhomogeneities in the neighbourhood of the ROI. As is revealed by the data of table 2.1, this is particularly brought out when Q is small.

Table 2.2 gives the average flow-rate computed using the MSD. It is observed that the average flow-rate arrived at this way is much smaller than that obtained using data from the pump used to circulate the liquid. We guess it is due to the mixing of dynamics in the capillary which renders the particles within suffer super-diffusive Brownian motion with α close to 2, Also, the micro-rotations introduced by the ultrasound forcing (which makes l^* smaller) are not considered in the present analysis. This discrepancy notwithstanding, the flow volume computed using $M(\tau)$ is closer to the truth than that obtained from $g_1(\tau)$.

Table 2.2. Flow rates (in μl s^{-1}) computed using $g_1(\tau)$ (Q_{g_1}) and $M(\tau)$ (Q_M) compared with the actual values used in the experiment. Reprinted with permission from [25]. (2015) Optical Society of America.

	19 μl s^{-1}	33 μl s^{-1}	47 μl s^{-1}	139 μl s^{-1}
Q_{g_1}	0.28	0.35	0.55	2.50
Q_M	2.47	3.03	4.29	20.17

To overcome the discrepancy seen above we did a 'no flow' experiment wherein the flow is nullified by the applied ultrasound force, currently applied in the opposite direction of the mean flow direction along the capillary. We have slowly varied the ultrasound force to ensure the drift in the computed $\langle \Delta r^2(\tau) \rangle$ is fully cancelled and the growth of $\langle \Delta r^2(\tau) \rangle$ with τ is proportional to τ, representing pure Brownian dynamics. With this we did two experiments; for flow volume computed using the applied ultrasound force to nullify it matched very well the values computed using alternative means. The average value from the experiment is: Q_{comp} 199.5 μl s^{-1} with the corresponding true value at Q_{true} 200 μl s^{-1}. The match is seen to be reasonable.

2.5 Conclusions

2.5.1 DWS in an inhomogeneous object

We have successfully extended DWS to cover objects with inhomogeneity in mechanical stiffness leading to inhomogeneity in the dynamics. We have used focused ultrasound beams to mark out regions and measured the Brownian-motion-induced decay in amplitude autocorrelation of light from the decay in the modulation introduced by the ultrasound on $g_1(\tau)$. Using this, the MSD of Brownian particles in the ultrasound focal region is measured and from it the visco-elastic spectra. We have made use of a generalized Langevin equation (GLE) to model the dynamics of the Brownian particles in the marked out region. From this model the MSD is computed. The match between results from experiments and theoretical model is quite satisfactory for the case of a homogeneous object where ultrasound tagging is not required. But this is not so when ultrasound is introduced and the MSD evaluated from the decay of $M(\tau)$. The match is seen to be perfect when τ is small, i.e. early rise time, but deviates in the plateau region of the MSD versus τ curve. The early-time match and deviation in the plateau region are observed with not only ultrasound-assisted experimental results, but also the usual DWS experiments done on homogeneous phantoms without ultrasound. The reason (for the plateau mismatch) is primarily owing to the mixing of the stochastic dynamics with the ultrasound-induced deterministic dynamics. This mixing pushes up the plateau region in the experimental results which can be matched only with a modified GLE with a multiplicative noise term in the restoration term. This exercise in modelling is deferred to a later chapter, chapter 4. Since the rise time match is good we conclude that the local property extracted is in fact the spatial average in the ROI of the inhomogeneous phantom. The other reason for this discrepancy is the upset of thermal equilibrium caused by the insonification, which pushes up the MSD. This problem can be at least partly addressed by reducing the ultrasound forcing in the focal region.

2.5.2 Detection and estimation of flow in a hidden capillary

Here, we have extended the application of ultrasound-assisted DWS to detect another dynamic inhomogeneity hidden in a multiple-scattering medium: which is liquid flow in a capillary hidden in a tissue-like object. Once again our measurement is the decay in the modulation introduced in $g_1(\tau)$ by the periodic ultrasound forcing in the ROI. An assumption under which we worked is that the external force introduced does not drastically change the flow. In particular, we have demonstrated, through experiments and simulations that the ultrasound-assisted local measurement is better suited to detect and measure the flow in tubes hidden in a multiple-scattering medium, than the normal DWS without ultrasound. With the ultrasound forcing the Brownian particles become super diffusive with $\langle \Delta r^2(\tau) \rangle$ rising as τ^α with α slightly short of 2. The ultrasound-assisted DWS captured this over a larger range of τ than was possible without ultrasound. The estimated $\langle \Delta r^2(\tau) \rangle$ is also an order of magnitude closer to its true value (determined from the knowledge of flow rate). Direct fallout of these demonstrations is the possibility of their application in biomedical imaging for detecting blood-flow changes in capillaries hidden in tissues, unhindered by the presence of the surrounding tissue. We note that the quantitative flow-rate recovered is still quite below actual values. The reason for this error is the mixing of ultrasound-induced dynamics with the Brownian dynamics which renders the flow super-diffusive. On the modelling front what is lacking is a GLE which captures this non-local effect (and hence the super-diffusive flow). A suitably designed multiplicative noise which will modulate the restoration term in the GLE should answer this shortcoming.

References

[1] Einstein A 1905 On the movement of small particles suspended in stationary liquids required by molecular-kinetic theory of heat *Ann. Phys.* **17** 549–60

[2] von Smoluchowsky M 1906 On the kinetic theory of molecules and their suspensions *Ann. Phys.* **326** 756–80

[3] Gisler T and Weitz D A 1998 Tracer micro-rheology in complex fluids *Curr. Opin. Colloid Interface Sci.* **3** 586–92

[4] Mason T G, Gang H and Weitz D A 1997 Diffusing-wave spectroscopy measurement of visco-elasticity of complex fluids *J. Opt. Soc. Am.* A **14** 139–49

[5] Boas D A, Campbell L E and Yodh A G 1995 Scattering and imaging diffusing temporal field correlations *Phys. Rev. Lett.* **75** 1855–8

[6] Zhou C *et al* 2006 Diffuse correlation tomography of cerebral blood flow during cortical spreading depression in rat brain *Opt. Express* **14** 1125–44

[7] Skipeterov S E and Megelinskii I V 1998 Diffusing wave spectroscopy in randomly inhomogeneous media with spatially localized scatterer flows *J. Exp. Theor. Phys.* **86** 661–5

[8] Chandran R S *et al* 2014 Diffusing-wave spectroscopy in an inhomogeneous object: local visco-elastic spectra from ultrasound-assisted measurement of correlation decay arising from the ultrasound focal volume *Phys. Rev.* E **90** 012303

[9] Eringen A C 1976 *Continuum Physics* vol 4 (New York: Academic)
[10] Mindlin R D and Eshel N N 1968 On first strain-gradient theories in linear elasticity *Int. J. Solids. Struct.* **4** 109–24
[11] Sarkar S *et al* 2015 Internal noise-driven generalized Langevin equation from a nonlocal continuum model *Phys. Rev.* E **92** 022150
[12] Kubo R 1966 The fluctuation-dissipation theorem *Rep. Prog. Phys.* **29** 255–84
[13] Kloeden P E *et al* 1994 *Numerical Solution of SDE through Computer Experiments* (Berlin: Springer)
[14] Priestley M 1967 Power spectral analysis of non-stationary random processes *J. Sound Vib.* **6** 86–97
[15] Cramer H 1961 On some classes of non-stationary stochastic processes *4th Berkley Symp Math. Statistics and Prob.* vol 2 (Berkeley, CA: University of California Press) pp 57–78
[16] Schtzel K and Drevel M 1988 Photon-correlation measurements at large lag times: improving statistical accuracy *J. Mod. Opt.* **35** 711–8
[17] Magatti T and Ferri F 2001 Fast multi-tau real-time software correlator for dynamic light scattering *Appl. Opt.* **40** 4011–21
[18] Technical Note No. TPHO9001E04 2005 Photon-counting using photomultiplier tube (Japan: Hamamtsu Photonics KK Electron Tube Division)
[19] Devi C U *et al* 2010 Design, fabrication and characterization of a tissue-equivalent phantom for optical elastography *J. Biomed. Opt.* **10** 44020-1–120-10
[20] Scheffold F *et al* 2003 Diffusing wave spectroscopy of non-ergodic media *Phys. Rev.* E **63** 061404
[21] Bicout D *et al* 1991 Dynamical correlations in multiple-scattering in Laminar flow *J. Phys. I France* **1** 471–91
[22] Heckmeier M *et al* 1997 Imaging dynamic heterogeneities in multiple-scattering media *J. Opt. Soc. Am.* A **14** 185–91
[23] Bicout D and Maret G 1994 Multiple light-scattering in Taylor–Couette flow *Physica* A **210** 87–112
[24] Bohren C F and Huffmann D R 1983 *Absorption and Scattering of Light by Small Particles* (New York: Wiley)
[25] Chandran R S *et al* 2015 Detection and estimation of liquid flow through a pipe in a tissue-like object with ultrasound-assisted diffuse correlation spectroscopy *J. Opt. Soc. Am.* A **32** 1888–97
[26] Wu X L *et al* 1990 Diffusing wave spectroscopy in shear flow *J. Opt. Soc. Am.* B **7** 15–20
[27] Han Y *et al* 2006 Brownian motion of an ellipsoid *Science* **314** 626–30
[28] Latini M and Bernoff A J 2001 Transient anomalous diffusion in Poiseuille flow *J. Fluid Mech.* **441** 391–411

Ram Mohan Vasu and Debasish Roy

Chapter 3

Mechanical property distribution from optical measurement of resonant ultrasound spectrum

3.1 Introduction

When a region in the object is remotely palpated by ultrasound forcing, and if the resultant frequency of the ultrasound forcing falls within the range of natural frequencies of vibration of the region, then these modes can be excited by the external force applied. By tuning of the frequency of the ultrasound one can observe this resonance of the focal region, the ROI, which also represents vibration of the scattering particles therein. With significant bearing on quantitative measurement of tissue properties, this resonance can be observed externally on the surface of the object. In the present context, coherent light is the carrier of this information, oscillation amplitude getting translated into a phase modulation of light, which manifests itself as a modulation on $g_1(\tau)$. (Coherent light need not play the role of carrier of acoustic information, as shown in chapter 4. Shear waves can modulate the compressional waves generated at the ROI, which can travel to the boundary carrying shear modulus information.) The ultrasound frequency at which this peak in modulation is detected can be taken as one of the natural modes of vibration of the ROI. Knowledge of the natural frequency (or frequencies) of vibration helps us compute the average shear-, or Young's modulus of the material of the ROI. In the first part of this chapter we demonstrate measurement of the most dominant mode of vibration of the ROI, and the recovery of Young's modulus from this measurement. In doing this, we analyze the response of the ROI to a periodic forcing by solving a linear momentum-balance equation describing the dynamics of the ROI. The parameters to be input are force distribution from acoustic radiation, geometrical shape of the ROI and the material properties such as density and shear modulus. The balance equation can be recast as an eigenvalue problem, the solution of which would give us a set of eigenvalues (natural frequencies) and eigenvectors (mode shapes). Since the elastic properties of the material enter into the eigenvalue

doi:10.1088/2053-2563/aae893ch3

equation, they can be recovered from the measured natural modes of the vibrating ROI. This formulation, which gives a frequency-measurement-based procedure for recovery of elastic properties, and its experimental validation form the first part of presentation in this chapter.

In the second part, we discuss the recovery of a spatially varying elasticity variation in the ROI when the object has an inhomogeneous material property distribution. In order to achieve a tomographic recovery of not just a constant value of, say, the Young's modulus which is the spatial mean for the ROI but a spatial variation of it, we need to enrich the measurement. Therefore, in this second study, we not only measure the resonant frequency, but also the variation of modulation depth, $M(\omega_f)$ as a function of ultrasound frequency, as ω_f is varied in the frequency bandwidth of interest which encompasses also the resonant frequencies. In [1] recovery of a related quantity, $p(\mathbf{r}) = \langle A(\mathbf{r})^2 \rangle_{(l^*)^3}$, where $A(\mathbf{r})$ is the amplitude of oscillation of the scattering centres, has been demonstrated from measurement of $M(\mathbf{r}, \omega_f)_{\mathbf{r} \in \partial\Omega}$ at a set of spatial locations on $\partial\Omega$ with ω_f kept fixed. We note that $p(\mathbf{r})$ is inversely related to the stiffness of the material in the ROI. Our motivation to extend this work to directly recover Young's modulus ($E(\mathbf{r})$) is two-fold: (1) we would like to rather recover $E(\mathbf{r})$ than a parameter only nonlinearly related to $E(\mathbf{r})$ and (2) in moving to an unknown, $E(\mathbf{r})$, which is independent of ω_f, we can vary ω_f and replenish data without a concomitant increase in the dimension of the unknown. This way, the ill-posedness of the inverse problem can be decreased. We also report demonstration of the recovery of $E(\mathbf{r})$ from experiments done using PVA phantoms.

3.2 Young's modulus recovery from measured natural frequency of vibration of ultrasound focal region

The procedure outlined here depends on the solution of the eigenvalue equation for the vibrating ROI (VROI) for its dominant modes; for then these can be compared with the measured natural frequency of vibration. The unknown elasticity parameter(s) can be extracted by bringing about a match between the computed and measured natural frequencies. (In what follows, we need only the average Young's modulus of the material in the VROI which is computed from just one, most dominant, natural frequency.) The main connection is established through a momentum-balance equation, whose solution gives the response of the VROI excited through a local periodic forcing. The momentum-balance equation is given by [2]

$$\rho \ddot{\mathbf{u}} = \nabla \cdot \sigma + F(\mathbf{x})\cos(2\pi\Delta f t) I_\Omega \tag{3.1}$$

Here ρ is the density of the material of the ROI, $\mathbf{u}$ is the vector displacement experienced by a representative scattering centre in the ROI, $\sigma = \sigma(\mathbf{x})$ is the Cauchy stress tensor and I_Ω is the indicator function of the VROI which undergoes oscillations because of the force $F(\mathbf{x})\cos(2\pi\Delta f t)$ from the ultrasound transducer in the ROI. The ROI is the intersection of the focal regions of a confocal dual-element transducer, where, as indicated in chapter 2, there is mixing of acoustic radiation at ω_f and $\omega_f + 2\pi\Delta f$ from the two regions. The low-frequency forcing, $F(\mathbf{x})\cos(2\pi\Delta f t)$, is of interest here, for it excites the shear modes of the ROI, which phase modulates coherent light interrogating it. In order to compute $\mathbf{u}$, or the dominant component of it (which is found to be along the common axis of the dual region ultrasound transducer) we need to input ρ, F and I_Ω in equation (3.1). We note that $\sigma(\mathbf{x})$ is in spatial coordinates $\mathbf{x}$ after deformation, which are unknown themselves. For small deformation we can approximate $F(\mathbf{x})$ by $F(\mathbf{X})$ where $\mathbf{X}$ represents coordinates before the body is deformed. Considering this, a more convenient way of writing equation (3.1) is through using the reference, material coordinates, $\mathbf{X}$. The new equation is given by [3]

$$\rho\ddot{\mathbf{U}} = \nabla \cdot \mathbf{P} + F(\mathbf{X})\cos(2\pi\Delta f t)I_\Omega \tag{3.2}$$

Here, $\mathbf{P}$ is the first Piola–Kirchhoff stress tensor which can be connected to σ [2] through the deformation gradient F and $\mathbf{U}(\mathbf{X})$ is the displacement vector with respect to the reference coordinates.

Density of the object used (in the simulations and later in experiments) is known, and is assumed invariant; F and I_Ω are computed using the procedure described below.

3.2.1 Force distribution in the focal region of the ultrasound transducer

We assume the force is significant only in the focal region of the transducer, and the displacement if any caused outside of it is neglected. The difference in the resonance frequencies of the two regions, Δf, is varied, in our case, from 10 Hz to 1 kHz. There is a low-frequency sinusoidal force at the ROI at Δf Hz because the acoustic force is proportional to intensity which in turn is proportional to square of the acoustic pressure. Denoting by $p_1(f, t)$ and $p_2(f + \Delta f, t)$ the pressure in the focal region owing to the two parts of the confocal transducer, they can be represented by $p_1(f, t) = p_{10}\cos(2\pi f t + \phi_1)$ and $p_2(f + \Delta f, t) = p_{20}\cos(2\pi(f + \Delta f)t + \phi_2)$. The average intensity $I(\mathbf{X})$ of the combined acoustic beams at the ROI is given by $I(\mathbf{X}) = \frac{1}{s}\frac{d\langle\tilde{\xi}(\mathbf{X})\rangle}{dt}$, where $\xi(\mathbf{X}) = \langle\tilde{\xi}(\mathbf{X})\rangle$ is the time-average over one cycle of the energy deposited at $\mathbf{X}$ and s is the area of intersection in the ROI we are dealing with. In [4] $\xi(\mathbf{X})$ is shown to be equal to $\xi(\mathbf{X}) = P_0 + P_1\cos(2\pi\Delta f t + \phi_2 - \phi_1)$ where P_0 and P_1 are shown to be dependent on p_{01}^2 and p_{02}^2 and T, the integration time-constant. Given these parameters, the radiation force experienced by the area around $\mathbf{X}$ is given by $F(\mathbf{X}) = \frac{2\alpha I(\mathbf{X})}{c}$, where α is the sound absorption coefficient and c the sound speed in the material of the ROI. This force, in our case, can be shown to be equal to

$$F(\mathbf{X}) = F_0 \sin(2\pi\Delta f t) \tag{3.3}$$

where $F_0 = \frac{4\alpha P_0}{Tcs}$.

3.2.2 Ultrasound-induced pressure distribution in the focal region

In order to find the pressure distribution in the focal region, the pressure at the surface of the transducer is propagated using a model which takes into account material nonlinearity (for the pressure amplitude can be large in the focal region), acoustic absorption and scattering in the medium. Such a model is the Westervelt equation [5]. Before we briefly describe the means of solving the above equation, we refer to an expression for $p_0(\mathbf{r})$, the pressure distribution on the transducer surface. This is given by $p_0^2 = \frac{2\rho c}{A} k_{\text{eff}}^2 \frac{V^2}{R}$. Here, A is the surface area of the transducer, k_{eff}^2 is the electrical-to-acoustic conversion coupling coefficient of the transducer (usually specified by the manufacturer) and R its equivalent resistance.

In order to transport p_0 on the surface of the transducer to its focal region (where the pressure can be very large), we employ the Westervelt equation, which is given by [5]

$$\nabla^2 p - \frac{1}{c^2}\frac{\partial^2 p}{\partial t^2} + \frac{\delta}{c^4}\frac{\partial^3 p}{\partial t^3} + \frac{\beta}{\rho c^4}\frac{\partial^2 p^2}{\partial t^2} = 0 \tag{3.4}$$

It is clearly seen that diffraction and linear propagation are represented by the first two terms. Thermo-viscous losses are accounted for by the third term and nonlinear effects by the fourth, where β is a constant dependent on the density of the medium and the average transducer pressure in the location. δ is sound diffusivity which has a say in acoustic absorption $\alpha = \frac{\delta\omega^2}{2c^3}$. We follow the method of [5] to solve equation (3.4) and expand the dependent variable using spheroidal coordinates. (The reader is referred to [5] for details.) We have put $\alpha = 2.5\text{dBcm}^{-1}$ and $\beta = 6.2$ in order to match values given in [6] for tissue-equivalent polymer material. We have computed pressure distribution in the focal volume with the transducer operating with one of the elements and then with both the regions energized. These pressure distributions are shown in figures 3.1–3.3, out of which figure 3.3 is a contour plot along the axial plane at $y = 0$.

3.2.3 Experimental verification of the shape of the focal volume

The shape of the focal volume in water is verified by focusing the composite transducer into a water bath and scanning the focal volume, one cross-sectional plane at a time, by a PVDF (polyvinylidene fluoride) transducer for pressure distribution. The experimental setup is the one used also for imaging and is described in section 3.2.6. In order to increase the spatial sample density the original acceptance area of the detector, approximately 7 mm^2, is masked down to 0.38 mm^2. The transducer axis being taken as the z-direction, at various depths along the axis the detector scans the x–y plane measuring a distribution of pressure values. The output from the measurement is voltages, which we take to be proportional to pressure which in turn is proportional to the force experienced by the particles in the ROI. With these assumptions, we have experimental measurement of force distributions in the ROI, in axial and radial planes which are represented in figures 3.4 and 3.5, respectively. They are close to the computed pressure distributions plotted in figures 3.1 and 3.2. Figure 3.6 is a contour plot of the measured voltage distribution in the axial x–z plane.

3.2.4 Determination of the boundary of the vibrating ROI

We arrive at the vibration amplitude of the scattering centres in the ROI by solving equation (3.1) (or equation (3.2)), which is the unforced linear momentum-balance

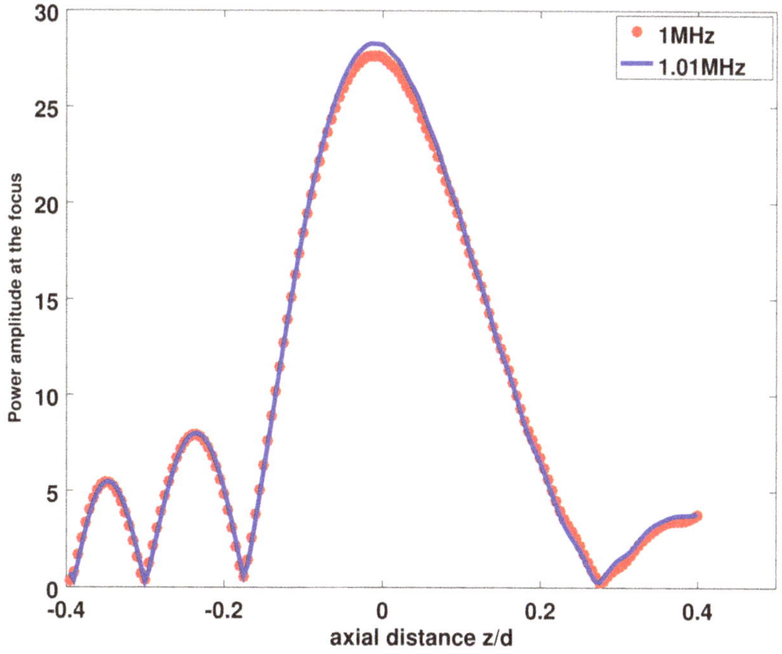

Figure 3.1. Computed pressure distribution along the transducer axis, i.e. (0, 0, z) (in arbitrary units). The origin is taken to be at the centre of the ROI. Here, d is the focal length of the ultrasound transducer. Adapted with permission from [2]. (2011) Optical Society of America.

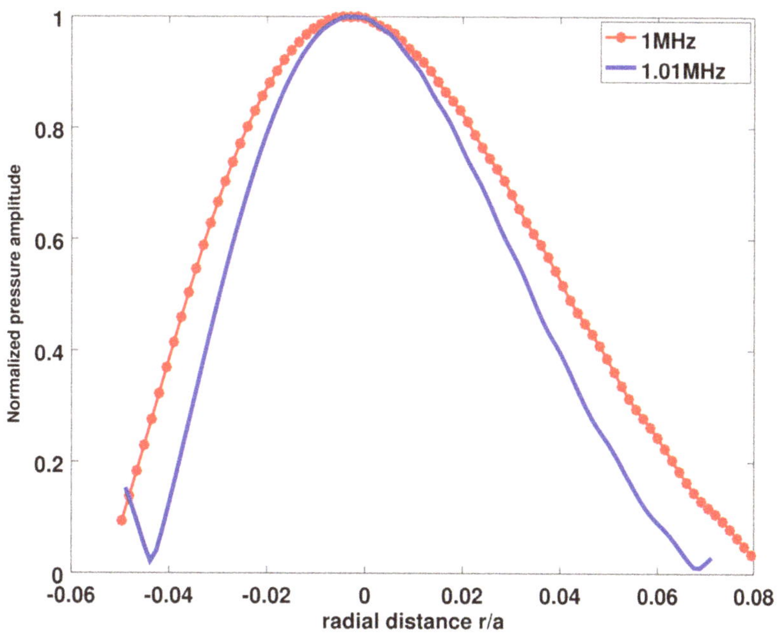

Figure 3.2. Normalized pressure distribution in a plane normal to the transducer axis i.e. (x, 0, 0). Here, a denotes radius of the ultrasound transducer. The origin is at the centre of the ROI. Adapted with permission from [2]. (2011) Optical Society of America.

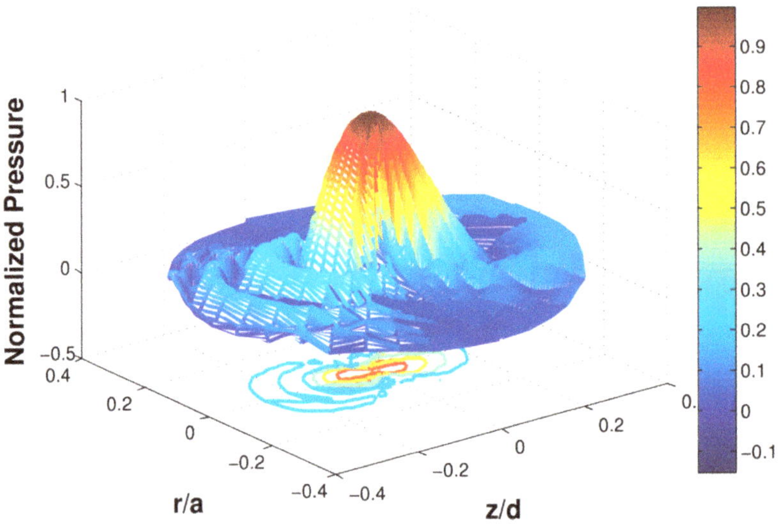

Figure 3.3. Mesh and contour plots of the pressure distribution in the $(x,0,z)$ plane. The ultrasound transducer has a focal length d and aperture radius a. The origin is at the centre of the ROI. Adapted with permission from [2]. (2011) Optical Society of America.

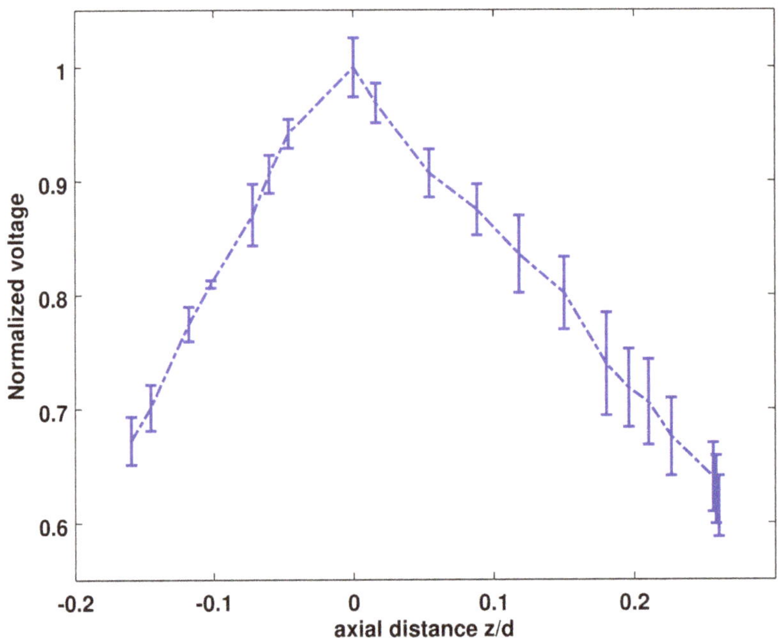

Figure 3.4. Experimentally measured voltage variation along the ultrasound transducer axis, which is a measure of the pressure distribution shown in figure 3.1. Parameter d is defined in figure 3.3. Adapted with permission from [2]. (2011) Optical Society of America.

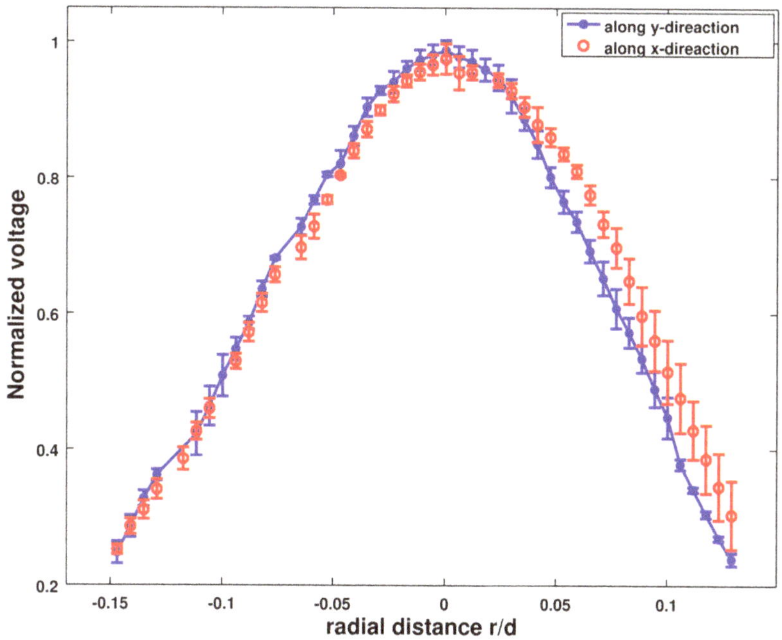

Figure 3.5. Experimentally measured voltage distribution normal to the ultrasound transducer axis. This verifies the pressure distribution plotted in figure 3.2. d is as defined in figure 3.3. Adapted with permission from [2]. (2011) Optical Society of America.

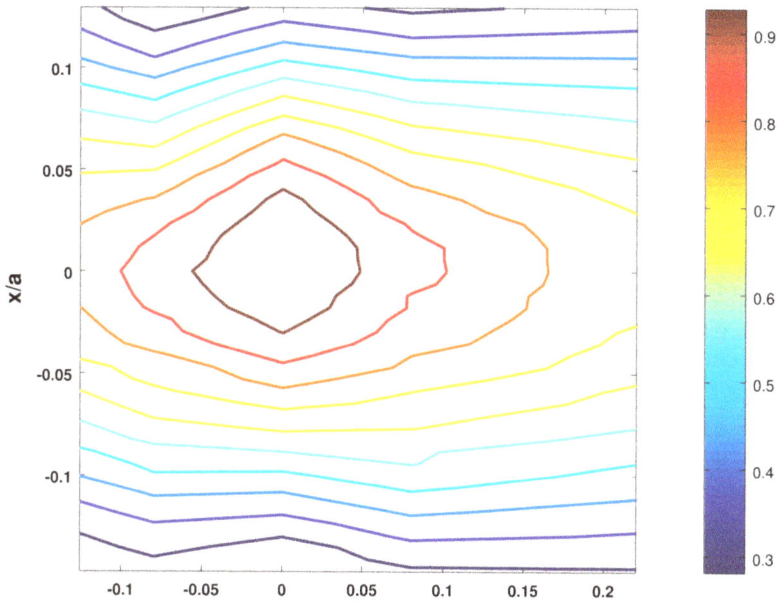

Figure 3.6. Contour plot of the experimentally measured and normalized voltage distribution in the x–z plane. Centre of the ROI is the origin. Adapted with permission from [2]. (2011) Optical Society of America.

equation for small deformation. We set it up as an eigenvalue problem and solve for eigenvalues (i.e. the weight of the natural frequencies) and the eigenmodes. For this, a commercial software package, ANSYS, is used. The result of the eigenanalysis is that we get the dominant vibrational frequencies owing to axial shear-deformation. To start the eigenanalysis, we choose the experimentally obtained $\partial\Omega$ (the closed boundary of Ω, the ROI), which is continuously refined as we proceed. Modified $\partial\Omega$ is obtained by correcting the boundary nodes from those identified, through solving the momentum-balance equation over one full cycle of forcing, as separating the vibrating ones from their non-vibrating complement. Pressure measurement in the confocal region revealed that pressure distribution from one region is not identical to that from the other, and one of them has a slight asymmetry (see figures 3.2 and 3.5). This difference is owing to a manufacturing defect of the confocal regions with the PZT coating.

The oscillations of the ROI under ultrasound forcing are characterized by a set of natural frequencies and mode-shapes (or eigenvectors) consisting of both compressional- and shear (dilational) modes. Of these, the compressional waves travel to the boundary without suffering appreciable attenuation, whereas shear waves suffer dissipation and are lost within the neighbourhood of the VROI. This shear wave attenuation is quite prominent at the higher frequency end of the eigenspectrum, which are the dilational modes generated by ultrasound forcing at $2f$ kHz. Even the low-frequency shear waves excited by the ultrasound forcing at the difference frequency attenuated quickly in the vicinity of the ROI. In chapter 4 we discuss modulation of the low-frequency compressional waves by shear modes, facilitating propagation of shear information to the boundary of the object through these compressional waves. Here we use coherent photons to carry shear deformation information as phase modulation of the interrogating coherent light, having suffered scattering from a multitude of particles within the VROI. At resonance, i.e. when the frequency of the externally applied sinusoidal forcing becomes one of the natural frequencies of the VROI, the oscillation suffered by the individual scattering centre will be the same as that suffered by the ROI. To see that, we express the forced vibration of the kth scattering centre (or, nodal point) in terms of its axial position $z_k(t)$; and $z_k(t)$ can be written using the modal basis function set and the particular solution of the balance equation under ultrasound forcing. That is, $z_k(t) = \sum_{l=1}^{n} c_k^l \Psi_l(t) + \gamma_k(t)$, where $\Psi_l(t)$ is the lth modal basis function, $\{c_k^l\}$ are the expansion coefficients, $\gamma_k(t)$ is the particular solution under ultrasound forcing and n the number of degrees of freedom of the discretized VROI. When the frequency of ultrasound forcing coincides with one of the natural frequencies, say ω_p, then $\gamma_k(t)$ becomes $\tilde{c}_k^p \Psi_p(t)$. Therefore, at resonance this term will dominate all the other terms in the expression for $z_k(t)$. Thus the ROI will oscillate with the same frequency as the individual scattering centres.

3.2.5 Transport of information from the ROI using coherent light

Transport of photons through a multiple-scattering object is already dealt with in chapter 2, section 2.3.2.1. In the present simulations we have also used Monte Carlo simulation to compute the ensemble average of phase, and its fluctuations, of photons reaching the detector. Monte Carlo simulation is computation intensive, but can accurately model photon transport in regions where the scattering coefficient is low, and also near the source and detector, where the diffusion approximation is invalid. The aim of the simulation is to compute $G^{\delta}(\mathbf{r}, \tau)$, the perturbation in amplitude autocorrelation introduced by ultrasound, and the dominant Fourier spectral components in it. The perturbation is a periodic modulation on the original, decaying background autocorrelation. The modulation depth is ascertained from the power spectral component at Δf Hz after the 'dc pedestal' is subtracted. Further details of signal extraction are available in [7].

In order to compute $G^{\delta}(\mathbf{r}, \tau)$, one can either use equations (2.19) and (2.20), in conjunction with equation (2.14), or use Monte Carlo simulation to propagate a large number of photons through the insonified object. The perturbation brought about by the ultrasound forcing in the ROI gives rise to phase fluctuations denoted by $\Delta\varphi$. Associated with a photon-path of length s we can have a phase fluctuation $\Delta\varphi^{s} = \Delta\varphi_{n}^{s} + \Delta\varphi_{d}^{s}$, with contributions from refractive index fluctuation and movement of scattering centres owing to the externally applied force. Considering tissue-like material used here nearly incompressible, we neglect refractive index fluctuation and therefore drop $\Delta\varphi_{n}^{s}$. The fluctuation in amplitude autocorrelation of light following path of length s is given by [8] $G_{s}^{\delta}(\mathbf{r}, \tau) = \exp[-\langle(\Delta\varphi)^{2}\rangle/2]$. Then $G^{\delta}(\mathbf{r}, \tau) = \int_{s} p(s)\, G_{s}^{\delta}(\mathbf{r}, \tau)ds$, where $p(s)ds$ is the probability that the length of photon path is between s and $s + ds$.

3.2.6 Experiments

3.2.6.1 Fabrication of tissue-mimicking phantoms with tailored properties

We have used PVA phantoms in this set of experiments. The recipe for preparation is given in [9] which also helps us tailor the storage modulus and scattering coefficient (within limits) of the fabricated slabs. By varying the so-called 'freeze-thaw cycles' [9], we could have storage modulus E varying from 11 to 97 kPa and μ_{s} from 1.4 to 4.5 mm^{-1}. Further increase of μ_{s} can be had through adding an appropriate quantity of polystyrene beads of diameter 0.1 μm. To characterize the slab for E a dynamic mechanical analyzer (DMA) was used; for μ_{s} an indirect method is employed, which uses the reverse Monte Carlo procedure [9]. The three slabs fabricated had storage modulus of 11, 45, and 58 kPa with a background scattering coefficient of 8.14 mm^{-1}.

3.2.6.2 Measurement of autocorrelation and the modulation depth

We first describe the experimental setup, which is not quite different from the one described in chapter 2. The schematic diagram in figure 3.7 describes the setup. A He–Ne laser (marked L, a 17 mW laser from Thorlabs, HRR 170) illuminates the PVA slab, P, which is also insonified by a dual-beam, confocal transducer, marked UST. The two portions of the UST are driven by an ultra-stable, dual-channel function generator (DCFG, Tektronix, model AFG 3022B) in conjunction with a power amplifier (PA). One of the portions is made to oscillate at 1 MHz and the

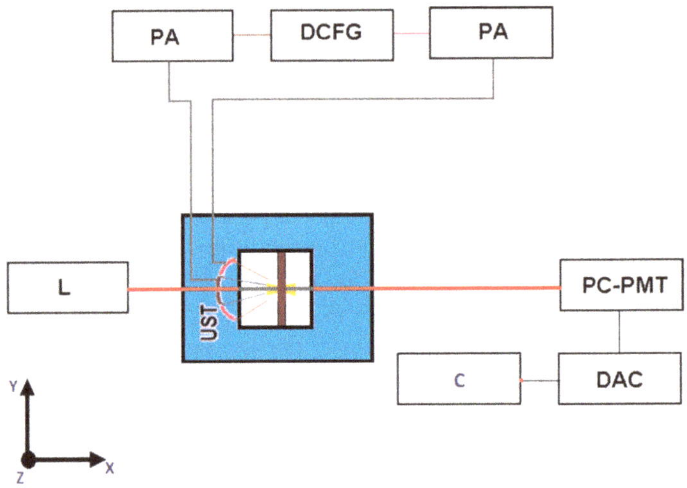

Figure 3.7. Schematic diagram of the experimental setup. A two-region, confocal ultrasound transducer (UST) insonifies a region in the object which is illuminated by the unexpanded beam from a laser L. Scattered light from the other end is picked-up and given to the detector, a photon-counting photomultiplier tube (PC-PMT). The output from the detector is given to the digital correlator (DAC) and further processed by the computer, C. The UST is driven by an ultra-stable, dual-channel function generator (DCFG) after power amplification. Reprinted with permission from [10]. (2015) Optical Society of America.

other at 1 MHz + Δf Hz. The acoustic beams at these frequencies mix at the common focal volume (the ROI) inside the object. Light from the laser with intensity regulated with the help of a neutral-density filter gains access to the slab, containing the ROI, through a hole in the transducer as shown. A photon-counting photomultiplier tube, PC-PMT, Hamamatsu H7360-03, detects the photons exiting the slab and water bath in which the object is immersed for acoustic impedance matching. Towards this, intensity fluctuation from a single speckle is first captured, with precise alignment in a single-mode fibre and is input to the PC-PMT. The detected voltage, after proper signal conditioning, is given to a hardwired correlator, the same used in the experiments of the earlier chapter, DAC Flex 021d from correlator.com. The correlator can give a minimum delay time of up to 12.5 ns. The output from the correlator, which is $g_2(\tau)$, is processed further in a computer. Processing is for extracting the modulation overriding it through power-spectral analysis. The two regions of the UST have a resonance frequency of 1 MHz, and we drive one of the regions Δf Hz away from this resonance. Owing to the presence of oscillations at this low beat frequency, and the fact that visco-elastic material at the ROI has its natural frequencies within a KHz, a low-frequency modulation is clearly seen on $g_2(\tau)$. We note in passing that the frequency stability of the driving function generator is of paramount importance in the experiments because our ability to generate a stable acoustic wave at Δf Hz critically depends on this.

In the experiments we use three PVA slabs (dimension 6 cm × 6 cm × 20 cm) of increasing E, namely 11, 45 and 58 kPa. Light intensity is gathered over 1200 s and $g_2(\tau)$ is computed through the autocorrelator. For any slab, $g_2(\tau)$ is found from the average of 2000 specimens computed. From $g_2(\tau)$ (measured for a particular Δf), the power-spectral amplitude at Δf Hz is found, which is designated $M^e(\Delta f)$ (experimentally measured modulation depth at Δf Hz). $M^e(\Delta f)$ plotted against Δf for the slab with $E = 11$ kPa is shown in figure 3.8. Also shown is the modulation depth obtained from the computed $g_2(\tau)$ ($M^c(\Delta f)$). It is observed that $M^c(\Delta f)$ closely follows $M^e(\Delta f)$, especially near the frequency where it peaks. This peak is evidence of the resonance of the VROI, here observed at ~70 Hz and measured at the boundary from the interrogating light carrying this information. Figures 3.9 and 3.10 are similar plots of modulation depth against frequency for slabs with $E = 45$ kPa and $E = 58$ kPa, respectively. We observe that resonant frequency observed in these plots increases from 70 Hz, to 140 Hz and then to 250 Hz.

We would like to recover E of the material of the slab in the ROI from the measured resonant peaks (we denote it by Δf_r). (Since the recovery is from only Δf_r data-model mismatch at other regions of the plots of figures 3.8–3.10 will not affect the accuracy of the recovered modulus of elasticity.) Towards this we use the method of bisection. For a guess of two values of Young's modulus, say, E_1 and E_2, such that the unknown E_0 falls in between these values, we first compute the resonant modes Δf_1 and Δf_2. Depending upon whether $\Delta f_m = \frac{(\Delta f_1 + \Delta f_2)}{2}$ and Δf_1 or Δf_2 contains Δf_r the search interval for E is modified to be from E_m to E_1 or E_2. Consequently the new search interval has become smaller. These steps are repeated until the computed resonant frequencies for edges of the interval match within a tolerance specified. The

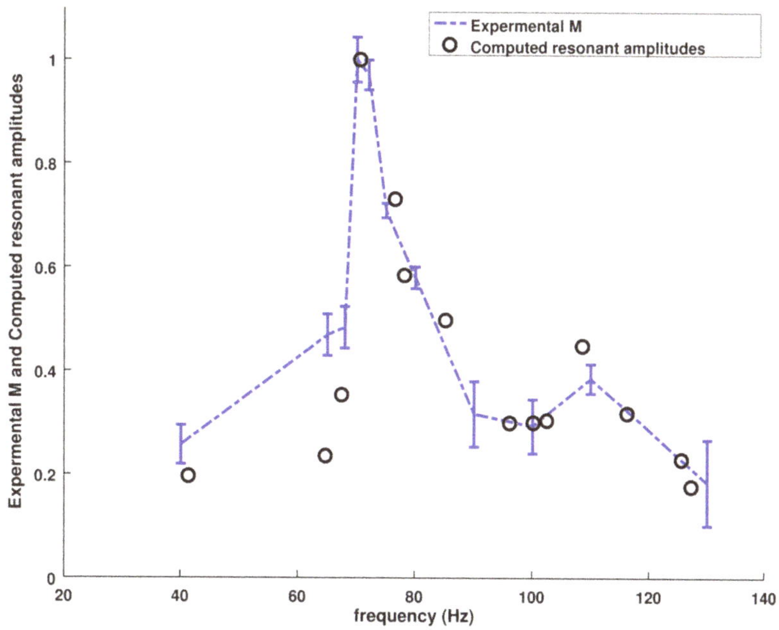

Figure 3.8. Variation of the experimentally measured modulation depth ($M^e(\Delta f)$) versus frequency for 11 kPa phantom and its comparison with the corresponding variation simulated. Both the plots are normalized by their maximum value observed at Δf_r, which for this phantom is 70 Hz. Adapted with permission from [2]. (2011) Optical Society of America.

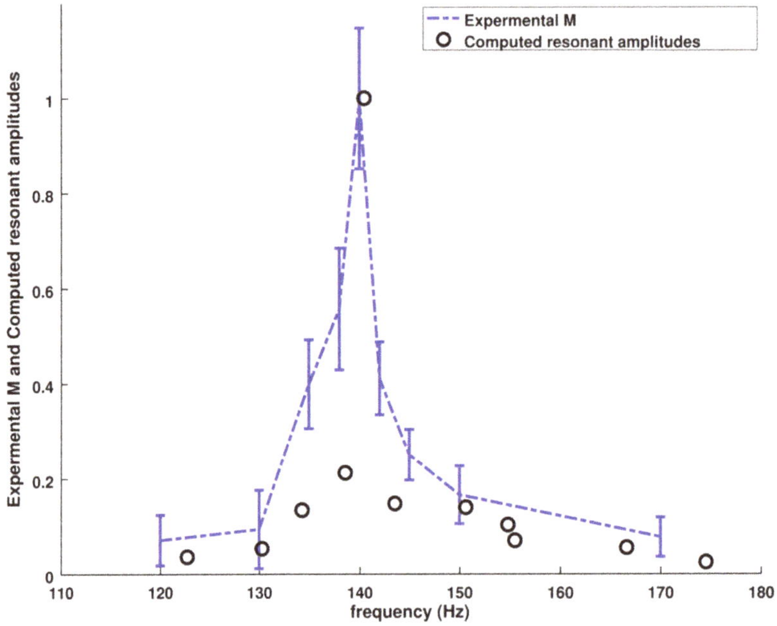

Figure 3.9. Same plots as those in figure 3.8, but for a phantom with $E = 45$ kPa. Adapted with permission from [2]. (2011) Optical Society of America.

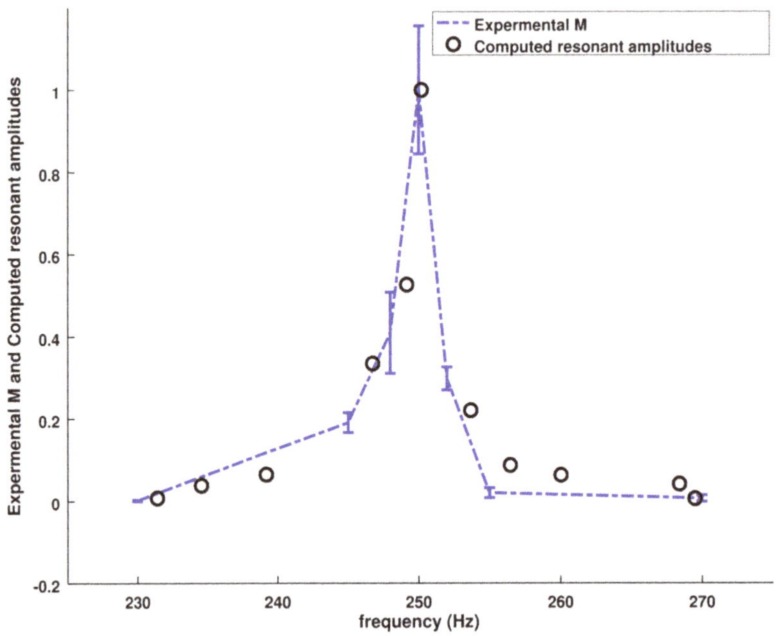

Figure 3.10. Same plots as those in figure 3.8, but for a phantom with $E = 58$ kPa. Adapted with permission from [2]. (2011) Optical Society of America.

unknown E values recovered for the three measured Δf_r's are 11.4, 44,8 and 58.7 kPa which match quite well the values measured for phantom samples using a rheometer.

3.3 Recovery of elasticity distribution in an inhomogeneous object

In this second part, both the theoretical basis and the experimental demonstration of recovery of inhomogeneous Young's modulus variation within the ROI are described. The method involves data collection, a sufficient set of 'orthogonal' data, from which the property distribution is recovered, as is done in any tomographic imaging, using the model(s) that connect data to the property. Variation of $M(\Delta f)$ when the detector moves around the ROI should give sufficient data for the recovery of $E(\mathbf{r})$. Since the alignment of the single-mode fibre to just one speckle to maximize signal-to-noise ratio (SNR), at a number of locations, is cumbersome, making the entire experiment unwieldy, we have made use of the additional parameter on which M depends, which is Δf, to gather a sufficiently large set of data. This data we gathered from just two detector locations.

The model connecting data to a parameter related to $E(\mathbf{r})$ is equation (2.19) of chapter 2, which is also given in equation (2.8) of [10]. This is a linearized perturbation equation connecting $G^\delta(\mathbf{r}, \tau)$ to $p(\mathbf{r}, \tau)$ after dropping the term containing the product of $G^\delta(\mathbf{r}, \tau)$ and $p(\mathbf{r}, \tau)$. We repeat this equation here for convenience:

$$\begin{aligned} &-\nabla \cdot \kappa \nabla G^\delta(\mathbf{r}, \tau) + \left(\mu_a + (1/3)k_0^2\mu_s'[\langle \Delta r^2(\mathbf{r}, \tau)\rangle]_b\right)G^\delta(\mathbf{r}, \tau) = \\ &-\left(\left(\sin^2\left(\tfrac{\omega_f \tau}{2}\right)p(\mathbf{r}, \tau) + (1/3)k_0^2[\langle \Delta r^2(\mathbf{r}, \tau)\rangle]_{in}\right)\chi_I \mu_s'\right)G(\mathbf{r}, \tau) \end{aligned} \tag{3.5}$$

It comes with the boundary condition $G^\delta(\mathbf{r}, \tau) + \kappa\frac{\partial G^\delta(\mathbf{r}, \tau)}{\partial \mathbf{n}} = 0,\ \mathbf{r} \in \partial\Omega$ (equation (2.20) of chapter 2).

We note that $p(\mathbf{r})$, though it can be connected to $E(\mathbf{r})$, is not the parameter distribution we want to recover. (In fact $p(\mathbf{r})$ as a parameter representing stiffness has been recovered using UMOT measurement, as in [1].) Therefore, our first motivation to directly recover Young's or shear modulus is that it is accepted as a mechanical property with direct correlation to the origin and progression of a tumour. The second motivation can be put forth with the help of an analogy taken from diffuse optical tomography (DOT). In a recent extension of DOT, known as the multi-spectral DOT, wavelength of light (ρ) is varied to enrich the fluence dataset and to recover the chromophores concentration directly [11]. The authors here have bypassed the usual two-step procedure wherein first the set of $\{\mu_a, \rho_j\}$ is reconstructed and concentration of haemoglobin in its oxygenated and deoxygenated states, as well as lipids and water are recovered from the absorption coefficients through inversion of an algebraic equation connecting them to the concentrations. Since the concentration of chromophores does not vary with ρ, with an enhanced dataset a considerable reduction in the ill-posedness of the DOT parameter recovery problem is achieved. This reduction was not possible in the two-step inversion, since dimension of the unknown μ_a increased with ρ.

Analogous to the spectral DOT, we introduce a spectral parameter in UMOT, which is the frequency difference between the two ultrasound transducers used in the experiments (Δf) to create the ROI by mixing the two focused acoustic beams. If we used $p(\mathbf{r})$ as the unknown in the first step of a two-step algorithm to recover $E(\mathbf{r})$ we gain no advantage in making the inverse problem well-posed; this is because $p(\mathbf{r})$ depends on Δf. Therefore, we make $E(\mathbf{r})$ our unknown, which is independent of Δf, through mapping $E(\mathbf{r})$ to $p(\mathbf{r})$ through a PDE, and a measurement equation. The PDE is the momentum-balance equation for the VROI, which connects $E(\mathbf{r})$ to $A(\mathbf{r})$, and the measurement equation is $p(\mathbf{r}) = \langle A(\mathbf{r})^2 \rangle_{(l^*)^3}$. With the introduction of the frequency Δf, we can also simplify data collection procedure by dispensing with detectors scouting many points on the object, and devising a scheme with one, or two, detectors collecting data over scan of frequencies covering the expected resonant modes of the ROI. As mentioned earlier, and demonstrated in the experiments in section 3.3.2, this rendered the data collection easier.

3.3.1 Theoretical formulation of the direct reconstruction problem

The idea of a direct recovery of $E(\mathbf{r})$ was considered advantageous following reports of direct recovery of chromophore concentration from spectral DOT data reported in [11, 12]. Even though the method followed here takes a different route. Like in many parameter recovery problems, we solve a mean-square error minimization problem with a regularization term added in the error functional to smoothen the reconstruction. The error functional is

$$\Theta(E) = \frac{1}{2} \parallel M^c(E) - M^e(E) \parallel^2_{L^2(\partial\Omega)} + \frac{\beta}{2} \parallel E \parallel^2_{L^2(\Omega_f)} \tag{3.6}$$

which is minimized with respect to $E \in L^\infty(\Omega_f)$. Here, β is an appropriate regularization parameter, and Ω_f the domain of the ROI with nonzero displacement. In equation (3.6) $M^e(E)$ is the experimentally measured modulation depth. However, $M^c(E)$ is the discretized version of a composite operator which taking E (along with other known properties of the object) as input, outputs the modulation depth. This composition has two PDEs and two measurement operators, out of which the PDE describing correlation diffusion (with which we can compute $G^\delta(\mathbf{r}, \tau)$ or $G(\mathbf{r}, \tau) + G^\delta(\mathbf{r}, \tau)$) is already described (equation (3.5)) and the measurement operator which acts on $G(\mathbf{r}, \tau) + G^\delta(\mathbf{r}, \tau)$ and computes modulation depth is a Fourier transform which helps us arrive at the power-spectral amplitude at the current Δf given by

$$M(r, p, \Delta f) = \left| \int_0^\infty (G + G^\delta)(\mathbf{r}, \tau) \exp\left[j\omega\tau \right] d\tau \right|_{\omega = 2\pi\Delta f} \tag{3.7}$$

The discretized version of equation (3.5) we write as

$$K(p)\mathbf{G}^\delta = \mathbf{s} \tag{3.8}$$

where $K(p)$ is the system matrix and $\mathbf{s}$ the source vector. The second measurement operator which takes the amplitude of vibration of the particles in the ROI ($A(\mathbf{r})$) as input and arrives at $p(\mathbf{r})$ has already been described. But the PDE connecting $E(\mathbf{r})$ to

$A(\mathbf{r})$ has not yet been discussed. We introduce this PDE, the momentum-balance equation, in the next section.

3.3.2 Momentum-balance equation

The VROI, denoted by Ω_f, is the domain under consideration for deriving the momentum-balance equation. This region in the object has nonzero amplitude of vibration as a consequence of insonification. As in the first experiment, we use a two-region, confocal ultrasound transducer and the common intersecting focal region contains the VROI, which can be identified by the method described in section 3.2.4 (i.e. $\partial\Omega_f$, the internal Dirichlet boundary separating the region with movement from the rest). Towards computing the displacement field (in our case the amplitude of vibration, $A(\mathbf{r})$) we have adopted a plane-stress approach based on 2D linear elasticity theory to setup the momentum-balance equation. The Poisson's ratio of the material of the VROI is taken to be 0.495, implying that the material is almost incompressible. In response to the ultrasound forcing, with these assumptions, the VROI undergoes sinusoidal oscillations in steady-state with the same frequency as the forcing. Therefore, a mixed form of the governing equations is made use of [13], which in terms of the amplitude, $A_0(\mathbf{r})$, of the displacement vector field $A(\mathbf{r}, t) = (\mathbf{A}(\mathbf{r}, t), q(\mathbf{r}, t))^T$ and pressure field $q(\mathbf{r}, t)$ takes the form

$$\rho\Delta\omega^2 A_0 + \nabla \cdot \left[-qI + \frac{E}{2(1+\nu)}(\nabla A_0 + (\nabla A_0)^T\right] = F_0, \mathbf{r} \in \Omega_f \tag{3.9a}$$

$$\nabla \cdot A_0 = \frac{(1+\nu)(1-2\nu)}{E\nu}, \quad \mathbf{r} \in \Omega_f \tag{3.9b}$$

$$A_0 = 0, \quad \mathbf{r} \in \Delta\Omega_f \tag{3.9c}$$

Here, $\Delta\omega = 2\pi\Delta f$ and F_0 is the ultrasound forcing amplitude. Here, all the quantities are defined in the un-deformed co-ordinates.

We apply the finite element method to solve equation (3.9) using the mixed weak formulation, even when the material approaches the incompressible limit and σ approaches 0.5. Through the weak formulation we aim to find $\mathbf{A} = (A_0, q) \in \ H^1(\Omega_f) \times L^2(\Omega_f)/\mathbb{R}$ so as to satisfy:

$$B(w, A) = (\mathrm{w}, F_0), \quad \text{for every } w = (\mathrm{w}, \varphi) \in H_0^1(\Omega_f) \times L^2(\Omega_f) \tag{3.10}$$

Here, $(\mathrm{w}, F_0) = \int_{\Omega_f} \mathrm{w} \cdot F_0 d\Omega_f$ is the linear form, and $B(w, A)$ is the bilinear form defined as:

$$\begin{aligned} B(w, A) = & \int_{\Omega_f}\Big(\rho(\Delta\omega)^2\omega \cdot A_0 - [\nabla w + \nabla w^T] \\ & : \left[\frac{E}{4(1+\nu)}(\nabla A_0 + (\nabla A_0)^T)\right]\Big)d\Omega_f \\ & + \int_{\Omega_f}\left((\nabla.\, w)q + \varphi\nabla \cdot A_0 + \frac{\varphi q}{\alpha E}\right)d\Omega_f \end{aligned} \tag{3.11}$$

We use α in equation (3.11) with a large value so that the bulk modulus is much larger than E. Pressure q can be eliminated using equation (3.9*b*). FE discretization of equation (3.10) leads us to the following matrix-vector equation:

$$K^h A_0^h = F^h \tag{3.12}$$

Here, K^h is the stiffness matrix, A_0^h the column-vector containing nodal values of the amplitude of vibration and F^h is the force-vector. Equation (3.12) can be solved for A_0^h from which we can compute the nodal values of the parameter $p(\mathbf{r})$.

3.3.3 Estimation of Young's modulus distribution through minimizing the error functional

With this we can decompose the operator $M^c(E)$ as a composition between two operators $M_A(p)$ and $M_B(E)$, where $M_A(p) = M_1 \circ \mathcal{F}_A$, and $M_B(E) = M_2 \circ \mathcal{F}_{\mathcal{B}}$. Here, $\mathcal{F}_A$ is given by $G^\delta = \mathcal{F}_A(\mu_a, \kappa, p)$, with $\mathcal{F}_A$ representing the inversion of the matrix-vector equation (3.8). Moreover, M_1 is the measurement operator given in equation (3.7). Similarly, $\mathcal{F}_{\mathcal{B}}$ represents the inversion of equation (3.12), and M_2 the measurement operator given by $p(\mathbf{r}) = \langle A(\mathbf{r})^2 \rangle_{(l^*)^3}$.

To solve the minimization problem of equation (3.6) we use the Gauss–Newton (GN) scheme. The GN iteration for the above minimization is:

$$E^{i+1} = E^i - (H(E^i))^{-1} G(E^i) \tag{3.13}$$

Here, H is the Hessian and G the gradient of the error functional of equation (3.6), evaluated at the current estimate of E. The GN approximations of H and G are:

$$H(E)\delta E = (D\mathcal{F}^*(E) D\mathcal{F}(E) + \beta I)\delta E \tag{3.14}$$

$$G(E) = D\mathcal{F}^*(\mathcal{F}(E) - M^e) \tag{3.15}$$

Here, we have used $\mathcal{F}$ to denote the operator M^C and $D\mathcal{F}$ its Frechet derivative. Similarly $D\mathcal{F}^*$ is the Frechet derivative of the adjoint of $\mathcal{F}$. The finite-dimensional equivalent of the Frechet derivative, the Jacobian matrix, has derivatives as its elements; typically, $\frac{\partial \mathcal{F}_i}{\partial E_j}$ is the rate of change of measurement at node i with respect to Young's modulus at node j. This derivative is obtained by combining the derivatives of M_A with respect to p and p with respect to E. Using a single forward solution of Frechet derivative of the CDE (correlation diffusion equation, equation (3.5)) and its adjoint, and making use of the reciprocity relation which correlation diffusion obeys, one can arrive at a complete row of the Jacobian matrix. For details, the reader is referred to [1]. Similarly, for $\frac{\partial p_i}{\partial E_j}$, the momentum-balance equation (equation (3.9)) and its adjoint [13] can be made use of. In the present application we have as many measurements as there are unknowns in Ω_f. Therefore, there is no computational advantage gained by using the adjoint scheme. Therefore, the perturbation scheme is made use of, which involves two forward solutions for any pair of unknown and measurement.

Whilst computing a typical element of the overall Jacobian matrix, one should note that a perturbation of the value of Young's modulus at jth node, E_j, produces perturbation in the value of p at all other nodes. Therefore, a typical $\frac{\partial \mathcal{F}_i}{\partial E_j}$ is evaluated as $\frac{\partial \mathcal{F}_i}{\partial E_j} = \frac{\partial \mathcal{F}_i}{\partial p_k}\frac{\partial p_k}{\partial E_j}$ and we need a full set of derivatives of the type $\frac{\partial p_k}{\partial E_j}$. These derivatives are obtained following the procedure noted earlier.

Once the derivatives are available, one can proceed with the GN iteration of equation (3.13). At each of the iteration the increment δE^i is computed making use of equations (3.14) and (3.15). A direct inversion of $H(E^i)$ is not attempted; instead, an optimization problem is set up from the perturbation equation and the optimum point is reached through a conjugate-gradient search scheme. The so-called 'outer' iteration is stopped when the error norm in the measurement domain reaches a preset small value and the current set of nodal values of E are taken as the solution.

3.3.4 Experiments using tissue-equivalent phantoms

3.3.4.1 Details of the phantom with inhomogeneity

The basic recipe for manufacturing of PVA phantom is already touched upon in section 3.2.5. The scattering coefficient of the fabricated slab is low compared to actual tissue samples, for we relied only on the trapped air bubbles within (during the 'cooling-thawing' cycles) to give it scattering property and did not introduce external scattering centres. The reduced scattering coefficient was measured and found to be approximately 8.0 cm^{-1}. Similarly the absorption coefficient of the slab (as fabricated, and without adding an absorber like India Ink) is found to be in the vicinity of 0.1 cm^{-1}. The mechanical properties of the two typical samples fabricated are also measured: $\rho = 1000 \text{ kg m}^{-3}$, $\nu = 0.490$ and $E = 11.3$ and 22.39 kPa. Slabs of dimension 3.5 cm × 3.5 cm × 3.5 cm are moulded. From these two composite slabs are made, first one by sandwiching a thin slice of Young's modulus 22.39 kPa (thickness 0.5 cm) between two slabs with $E = 11.3$ kPa, and the second by attaching two thin slices of $E = 22.39$ kPa at the ends of a slab of uniform Young's modulus of $E = 11.3$ kPa. The total thickness of the composite slab, in both the cases, was kept at 3.5 cm. Schematic diagrams representing the cross-sections of these slabs are shown in figure 3.11

3.3.4.2 Experimental setup and data collection

The experimental setup remains unchanged with respect to the one described in the first experiment here (figure 3.7), except for a provision to rotate the orientation of the object so that angle at which laser light enters the object can be turned by 90°. The new orientation of the object is shown schematically in figure 3.12. The composite object, as described earlier, is immersed in a water-bath for acoustic impedance matching. An unexpanded beam from a 15 mW He–Ne laser illuminates the object from one side. The difference between ultrasound drive frequencies (Δf) is varied from a few tens of Hz to 2 kHz. The common focal region of the confocal ultrasound transducer (at its thinnest waist) is approximately of hyperboloidal shape, with length and breadth, respectively, of 1.42 and 0.2 cm. With the help of

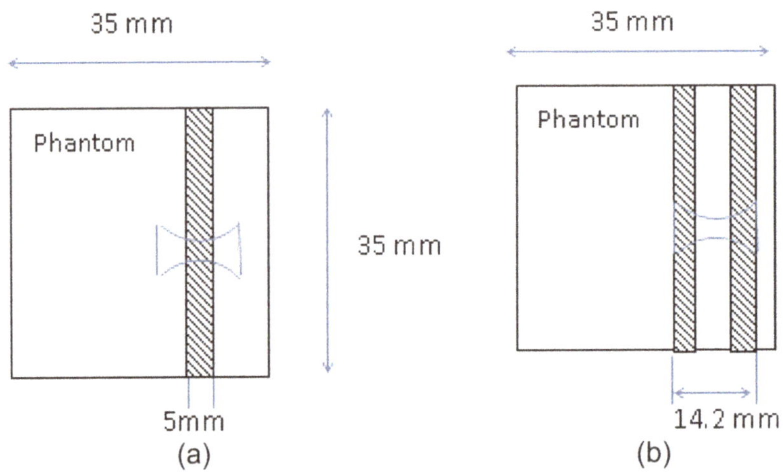

Figure 3.11. Cross-section of the composite PVA phantom. Here, the hatched region has an E of 22.39 kPa and the other 11 kPa: (a) object 1 and (b) object 2. Reprinted with permission from [10]. (2015) Optical Society of America.

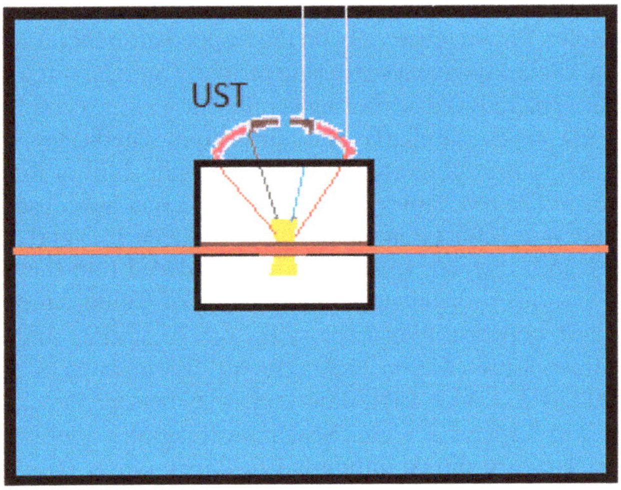

Figure 3.12. The object in figure 3.7 after it is rotated by 90° allowing us another 'view' of the object. Reprinted with permission from [10]. (2015) Optical Society of America.

translation stages on which the object is mounted, the common focal volume is ensured to intercept the region in the slab where the inhomogeneity in E is (figure 3.17). As in the first experiment, a single-mode fibre is used to capture just one speckle from the light exiting the bath from the opposite side to illumination; the intensity fluctuation is input to the PC-PMT. The output from the PC-PMT, after proper signal conditioning is given to the autocorrelator, from which we get $g_2(\tau)$. From $g_2(\tau)$, $|g_1(\tau)|$ is computed using Siegert relation [14]; the power-spectral amplitude at the selected Δf, after subtracting the 'dc pedestal', is the measurement

($M^e(E, \Delta f)$) denoted also by M_1. The sampling frequency of $g_2(\tau)$ output from the hardwired autocorrelator can be set depending on the maximum frequency of the signal we want to extract from it. Since the resonant modes of the ROI enclosing a visco-elastic medium seldom exceed 1–2 kHz, we have selected 100 μs as the sampling time used in the autocorrelator.

As mentioned earlier, we collect a large set of data, not by moving the source-detector combination around, but by varying Δf, i.e. $M^e(E, \Delta f_i)_{i=1,\dots n}$. This data set from a single source location was used to recover $E(\mathbf{r})$, but the data was found to be insufficient for a recovery which was quantitatively accurate enough. Therefore, in the present experiments, it was decided to collect data from one more detector location. Without disturbing the fibre alignment, the object on the translation-cum-rotation stage was rotated by 90° so that the illumination enters it from a different angle, as shown in figure 3.12. In this orientation we have collected a fresh set of data by varying Δf. The two sets together formed the input data for the GN algorithm for recovering $E(\mathbf{r})$.The data gathered using the two objects shown in figures 3.11(a) and (b) are in figures 3.13–3.16, when Δf is varied from 200 to 700 Hz. We note that there is considerable variation in modulation in the above frequency range, and it also encloses peaks which point towards resonant behaviour of the VROI. The step-size of frequency increase is selected depending on stability of the ultrasound source, which in our experiments was in the range of 2 Hz to 10 Hz.

Using properties of the object and the geometry of experimental setup, we have also generated data through numerical simulation. The extra information required for simulation is the size and shape of the ROI and the force applied therein at the

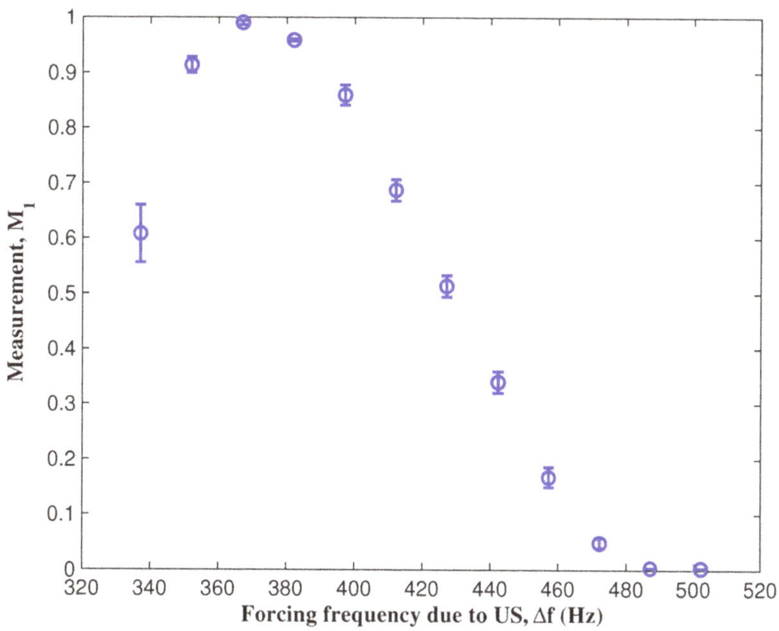

Figure 3.13. Variation of experimentally measured M_1 with ultrasound (US) beat-frequency for object 1 with illumination as in figure 3.7. Reprinted with permission from [10]. (2015) Optical Society of America.

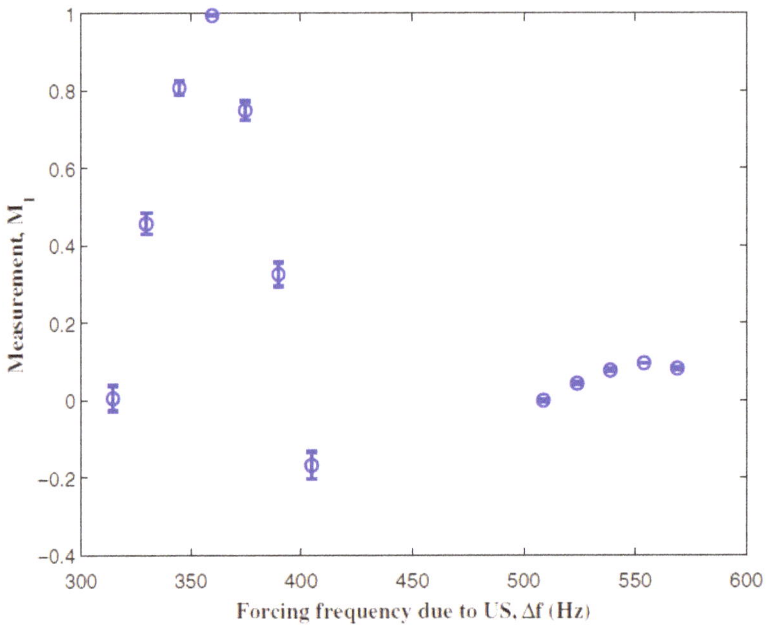

Figure 3.14. Same as figure 3.13, but for object 2. Reprinted with permission from [10]. (2015) Optical Society of America.

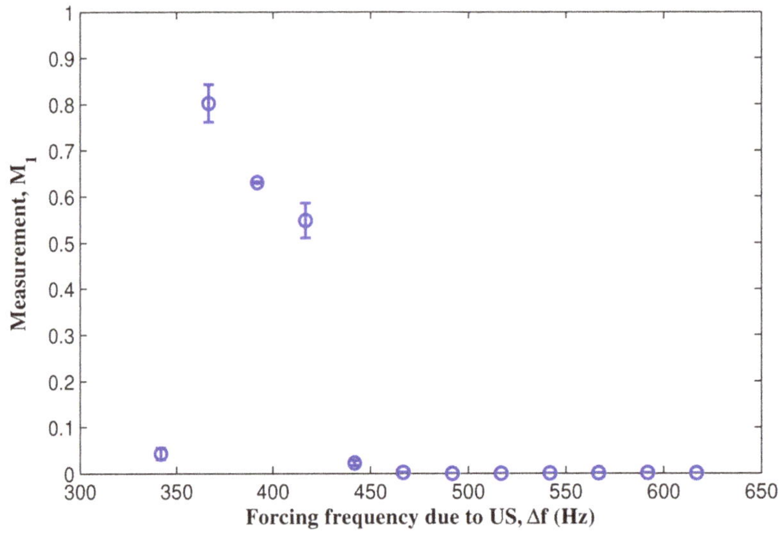

Figure 3.15. Variation of experimentally measured M_1 with ultrasound (US) beat-frequency for object 1 with illumination as in figure 3.12. Reprinted with permission from [10]. (2015) Optical Society of America.

ultrasound difference frequency. These are computed using the procedure described earlier in section 3.2.2. For generating data, both the CDE and the momentum-balance equation needed to be solved. In keeping with the experiments, data for two detector locations 90° apart are gathered.

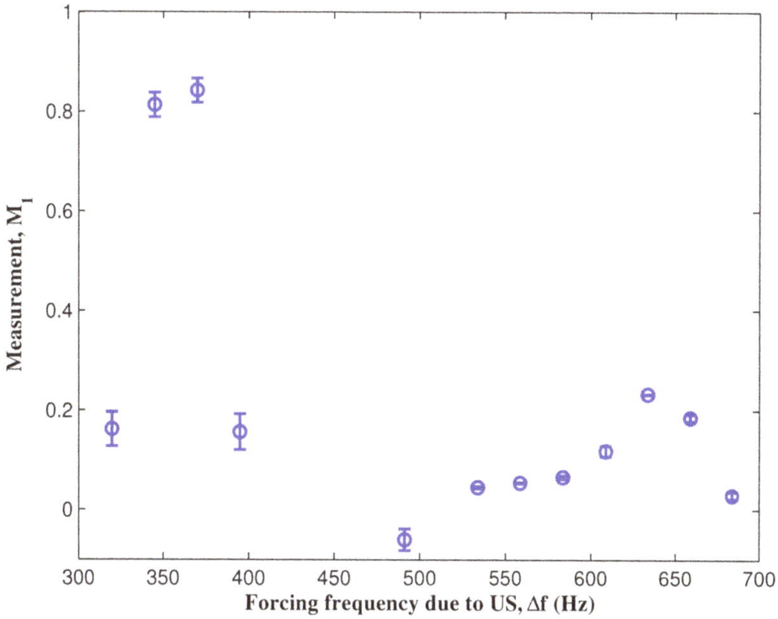

Figure 3.16. Same as figure 3.15, but for object 2. Reprinted with permission from [10]. (2015) Optical Society of America.

In all the numerical algorithms, both for generating data and for recovering $E(\mathbf{r})$ within the ROI, we have employed 2D objects which are cross-sections of the ROI containing the ultrasound transducer axis. We note that the plane-stress approximation is not necessarily valid across any cross-section of the ROI. However, to demonstrate the working of the algorithm for the recovery of inhomogeneities of $E(\mathbf{r})$ in a computationally expedient manner, both in location and quantitative accuracy, we chose to restrict ourselves to 2D objects within the plane-stress approximation. For generating data numerically, we have employed a finer mesh compared to that used for inversion. Here, fixing a square cross-section of the ROI of 3.5 cm × 3.5 cm, we have used an FE mesh with 13 168 triangular elements and 6740 nodes. To the computed data was added 2%–4% noise to arrive at $M^C(\mathbf{r}, \Delta f)_{\mathbf{r}\in\partial\Omega}$.

The data gathered experimentally, and generated through numerically solving the forward equations, are input to the iterative GN procedure discussed in section 3.3.3. The perturbation equation, equations (3.13)–(3.15), is inverted, using conjugate-gradient search scheme, to arrive at the update in $E^i(\mathbf{r})$, which is $\delta E^i(\mathbf{r})$. Since the recovery of $E(\mathbf{r})$ is a highly nonlinear problem, the perturbation equation itself is updated at the end of each iteration by re-computing the Jacobian and the measurement error. The regularization term βI (β is the regularization parameter and I the identity matrix) is employed in equation (3.14) where β is taken to be 3.2116×10^{-54} at the start of inner iteration, and reduced by a factor of 1.5 after each main iteration. The value of β needs to be arrived at separately for every data-set presented to the inversion routine. One empirical means of arriving at an appropriate starting value is through analyzing the diagonal elements of the sensitivity matrix (or, Jacobian) computed for setting up the perturbation equation.

3.3.4.3 Results and discussion

The results of iterative recovery are presented in two separate sets of figures, which correspond to input data from two different inhomogeneity distributions in the ROI shown in figures 3.17(a) and (b). By varying Δf we have gathered 63 different measurements from each location of the source–detector pair; putting them together we had a data set of dimension 126. Results of reconstruction of the Young's modulus distributions in the ROI for the two cases (grey-scale images) are shown in figures 3.18(a) and (b). Cross-sectional plots through the original object and those through reconstructions using experimental and simulated data are shown in figures 3.19(a) (for the object shown in figure 3.18(a)) and 3.19(b) (for the object shown in figure 3.18(b)). The behaviour of the data-domain mean-squared error is shown in figures 3.20(a) and (b) in respect of the two distributions being recovered. It is seen

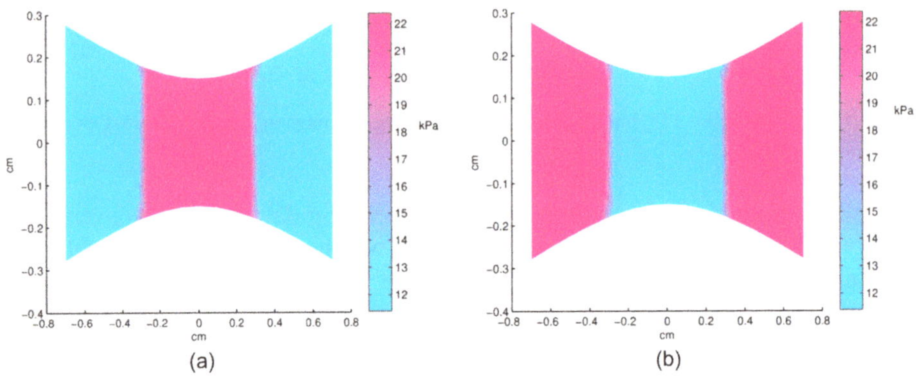

Figure 3.17. Young's modulus distribution in the inhomogeneous region of the composite object where the ultrasound common focal volume is: (a) object 1 and (b) object 2. Reprinted with permission from [10]. (2015) Optical Society of America.

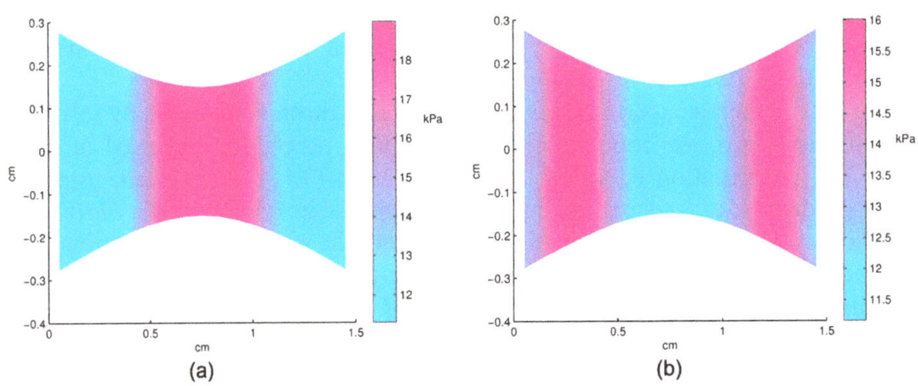

Figure 3.18. Reconstruction of $E(\mathbf{r})$ using Gauss–Newton algorithm: (a) for the original shown in figure 3.17(a) and (b) for the one shown in figure 3.17(b). Reprinted with permission from [10]. (2015) Optical Society of America.

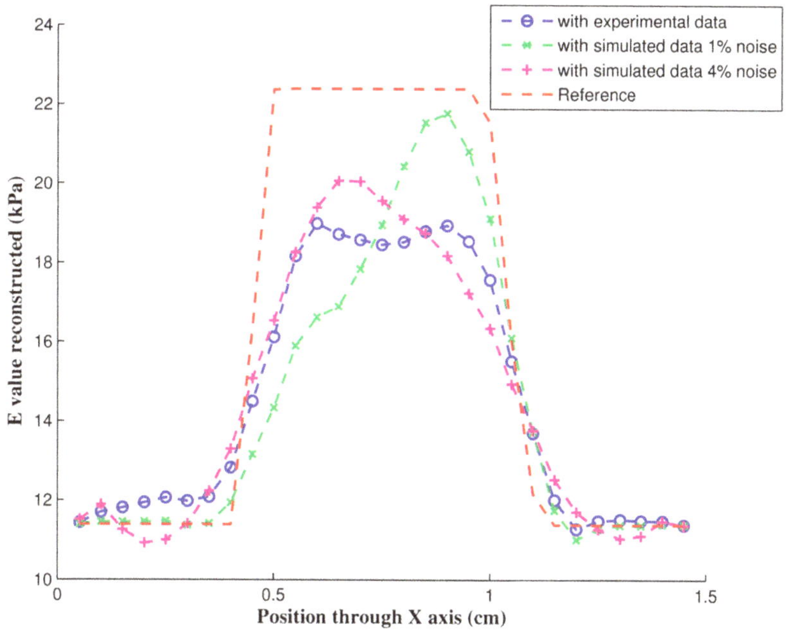

Figure 3.19a. Cross-sectional plots through the centre of the reconstructions from experimental and simulated data corresponding to object 1 shown in figure 3.17(a). The reference cross-section is from the original shown in figure 3.17(a). Reprinted with permission from [10]. (2015) Optical Society of America.

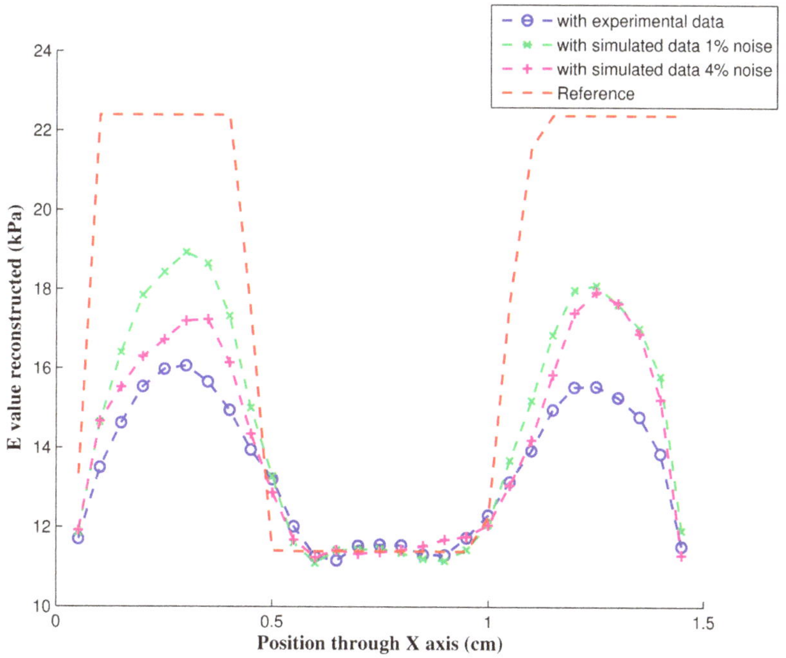

Figure 3.19b. Same as figure 3.19(a), but for object 2. Reprinted with permission from [10]. (2015) Optical Society of America.

that, for the first object, the error reduced to 2×10^{-24} in 11 iterations, whereas for the second object it reduced to 3×10^{-23} in six iterations. From these results it is evident that data collected through varying the ultrasound frequency, one is able to generate linearly independent data sets for faithful and accurate recovery of $E(\mathbf{r})$ from UMOT data. The advantage this holds for a practical realization of an imaging system for mapping mechanical property is that data collection is less cumbersome and therefore easily manageable.

3.3.5 Conclusions

In the first part of the work presented in this chapter we have demonstrated recovery of quantitative Young's modulus value corresponding to an average for the region selected by the ultrasound focal volume, the ROI. The data is the natural frequency of the vibrating ROI, which is carried to the boundary of the object with the help of coherent light used to interrogate the object. Since the basic measurement is frequency, the data is unaffected by variations in acoustic- or light intensity. Since modulation depth is embedded in background noise and hence difficult to measure in the usual UMOT experiments to recover $\mu_a(\mathbf{r})$, the measurements at resonance will enhance the SNR of the measured modulation depth, making recovery of optical parameters also more accurate.

In the second part of the work, we have extended the previous measurement to gather variation of modulation around the resonant peaks: i.e. we have the behaviour of M with Δf the ultrasound beat-frequency. Using this data set,

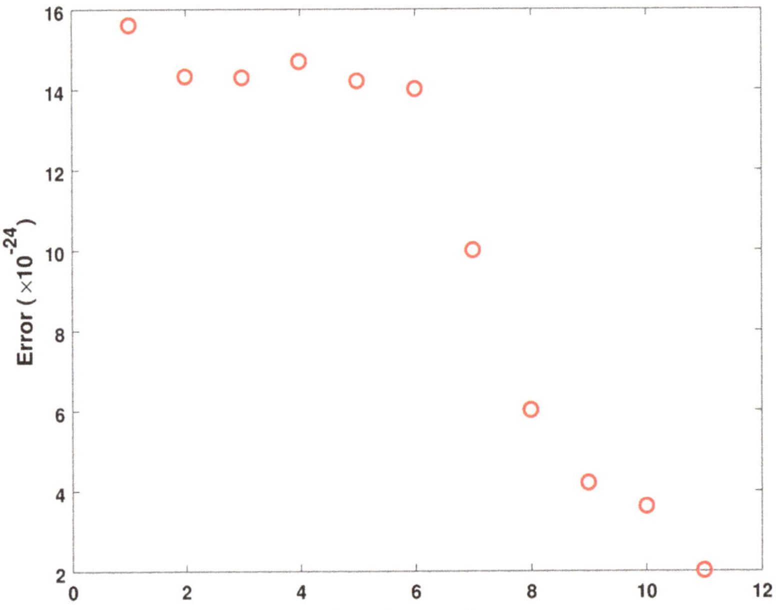

Figure 3.20a. Data-domain mean-square error versus iteration number when reconstructing the cross-section of object 1. Reprinted with permission from [10]. (2015) Optical Society of America.

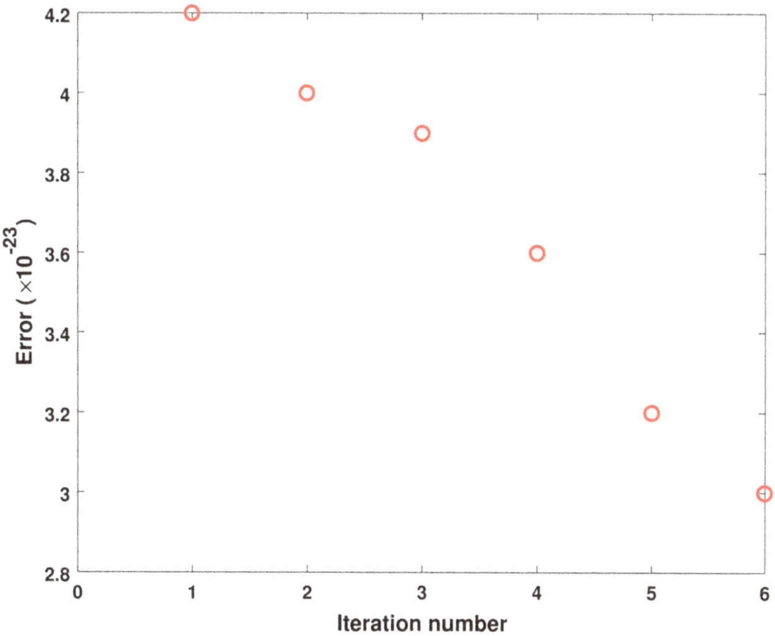

Figure 3.20b. Data-domain mean-square error versus iteration number when reconstructing the cross-section of object 2. Reprinted with permission from [10]. (2015) Optical Society of America.

inhomogeneous Young's modulus distribution in the ROI is recovered. Data gathering for this tomographic recovery was easy, because scanning of the acoustic frequency was less cumbersome compared to moving the alignment-sensitive 'gantry' around. Therefore, the simplified experiments introduced here have the potential to move the current experimental prototype to robust imaging instruments for medical diagnosis.

References

[1] Varma H M *et al* 2011 Ultrasound-modulated optical tomography: recovery of amplitude of vibration in the insonified region from boundary measurement of light correlation *J. Opt. Soc.* A **28** 2322–31

[2] Chandran R S *et al* 2011 Ultrasound-modulated optical tomography: Young's modulus of the insonified region from measurement of natural frequency of vibration *Opt. Express* **19** 22837–50

[3] Marsden J E and Hughes T J R 1993 *Mathematical Foundations of Elasticity* (New York: Dover)

[4] Konofagou E *et al* 2001 A focused ultrasound method for simultaneous diagnostic and therapeutic applications: a simulation study *Phys. Med. Biol.* **46** 2967–84

[5] Kamakura T *et al* 2000 Model equation for strongly focused finite-amplitude sound beams *J. Acoust. Soc. Am.* **107** 3035–46

[6] Duck F A 2002 Nonlinear acoustics in diagnostic ultrasound *Ultrasound Med. Biol.* **28** 1–18

[7] Mazumder D *et al* 2017 Quantitative vibro-acoustography of tissue-like objects by measurement of resonant modes *Phys. Med. Biol.* **62** 107–26

[8] Sakadzic S and Wang L V 2002 Ultrasonic modulation of multiply scattered coherent light: an analytical model for anisotropically scattering media *Phys. Rev.* E **66** 062603
[9] Usha Devi C *et al* 2005 Design, fabrication, and characterization of tissue-equivalent phantoms for optical elastography *J. Biomed. Opt.* **10** 044020
[10] Mohanan K P *et al* 2015 Ultrasound-modulated optical tomography: direct recovery of elasticity distribution from experimentally measured intensity autocorrelation *J. Opt. Soc. Am.* A **32** 955–63
[11] Srinivasn S *et al* 2005 Spectrally constrained chromophore and scattering near-infrared tomography provides quantitative and robust reconstruction *Appl. Opt.* **44** 1858–69
[12] Kim H M *et al* 2010 PDE-constrained multispectral imaging of tissue chromophores with equation of radiative transfer *Biomed. Opt. Express* **1** 812–24
[13] Oberai A A *et al* 2003 Solution of the inverse problem in elasticity imaging using the adjoint method *Inverse Probl.* **19** 297–313
[14] Berne B J and Pecora R 2000 *Dynamic Light Scattering* (New York: Dover)

IOP Publishing

Ultrasound-Mediated Imaging of Soft Materials

Ram Mohan Vasu and Debasish Roy

Chapter 4

Quantitative vibro-acoustography from measurement of modal frequencies: characterisation of isotropic and orthotropic tissue-like objects

4.1 Introduction

In chapter 3 we reported recovery Young's modulus in the ROI, first an average value from measurement of the dominant natural frequency of vibration of the ROI, then a distribution of Young's modulus from a fuller measure of spectrum of the VROI. Coherent light was used to probe the object, which would carry information on dynamics of the scattering centres in the ROI as phase modulation, which, in turn, modulates the decaying intensity (or amplitude) correlation of light measured outside the object. The first of these, which like resonant ultrasound spectroscopy (RUS), depends on experimental ability to measure resonant modes of vibration so that material properties of the region can be inverted. Our experience with the experiments of the previous chapter confirms the major difficulty of RUS [1] to measure high-frequency modes, which become too weak for detection as order increases. In jelly-like materials which form the object of our interest here, the background temperature-induced Brownian motion can be of large intensity which mixes with the deterministic dynamics introduced by ultrasound forcing. The Brownian power-spectrum has $\frac{1}{\omega^2}$ decay in frequency, and this multiplies the power-spectrum of the dynamics introduced by the ultrasound forcing [2]. Because of this interference the higher harmonics of the natural frequencies present in the VROI become all the more weak. The primary objective of the present work is to prove the efficacy of a light-based probe to amplify and measure many of the natural frequencies of vibration of the ROI. The probe is diffusing-wave spectroscopy (DWS) in the presence of ultrasound modulation in a selected region, which is the ROI. The growth of the MSD with

doi:10.1088/2053-2563/aae893ch4

lag-time, in a visco-elastic jelly-like object, obtained from the decay of autocorrelation of light, attains a plateau when the thermal energy in the system is moderate. The plateau indicates that the Brownian particles, modelled to be connected through springs and dashpots, are stretched to the maximum. The observation in [3] is that the introduction of the ultrasound forcing makes the MSD fluctuate in the plateau region, giving it a noisy appearance. It is also seen that the dynamics of the ROI is available in the fluctuations in the plateau: Fourier transform of the fluctuations reveals the natural frequencies and mode-shapes pertaining to the VROI. This amplification of the higher order modes, brought out only when probed by photons, is owing to a 'coherent' cooperative interaction between movement due to background noise and ultrasound-induced periodic (sinusoidal) forcing. This cooperative phenomenon, which shows a resonant-like behaviour with respect to noise intensity and also frequency of the sinusoid, has been well-researched in the past [4, 5] as stochastic resonance (SR), wherein energy is resonantly transferred from noise to an overriding weak sinusoid. We will return to the topic of noise-assisted enhancement of weak higher harmonics (or, modes) present in the collective dynamics of the VROI in section 4.4.1. The enhancement of sinusoids, brought about through the DWS probe, is also owing to the chains of oscillators which are in the diffusing light paths from which photons scatter and reach the detector. The collective behaviour of the chain of scattering centres in the ROI, modelled to be coupled to each other through visco-elasticity and driven by noise and weak sinusoids, exhibits a further enhancement of SNR as the coupling strengths are varied, showing a resonant behaviour against the coupling strength [6].

Before we take up the topic of the measurement tool that DWS provides, we discuss in section 4.2 vibro-acoustography. With this, we explain the passage of highly attenuating shear waves produced in the ROI through modulation of the compressional, vibro-acoustic (VA) wave, which is not attenuating. The VA wave is detected on the boundary of the object with the help of a fibre Bragg grating (FBG). It is shown that the VA amplitude is proportional to the local shear wave amplitude at the ROI, as the propagation equation for the compressional wave is driven by a source term which has the shear-wave amplitude as a factor. Measurement of the fundamental frequency of vibration of the VROI is experimentally demonstrated through the measured VA-wave amplitude and the ultrasound frequency at which it peaks (see section 4.3). It is also noticed that the higher harmonics are too weak to be detected through this means. In section 4.4, we discuss DWS, and how it can provide a measurement tool for a larger set of natural frequencies of the VROI. We further demonstrate the measurement of natural frequencies of vibration in pork tissue, assumed to be orthotropic in elastic behaviour, and recover all the nine components of the elasticity tensor from the measured set of frequencies. The concluding remarks are set out in section 4.5.

4.2 Quantitative vibro-acoustography and measurement of resonant modes of region of interest

As mentioned in chapter 1, study of generation of VA wave in the ROI gives a better insight into the dynamics of the ROI set in motion by ultrasound forcing and how local shear wave gets coupled to the propagating compressional wave. The main objective of elasticity modulus recovery remaining unaltered, the work presented in [7] by some of us tries to explain how the detected amplitude of the VA wave faithfully follows variation in the shear modulus in the ROI. Following [7], we present here a theoretical basis for the claim that natural frequencies vibration of the ROI can be obtained from the spectrum of the detected VA wave. Towards this, we derive the propagation equation for the VA wave which has a source term containing the shear displacement field produced locally within the ROI. Once the natural frequency, or frequencies, is inferred from the VA spectrum they are inverted for the average elastic modulus of the material in the ROI.

4.2.1 Propagation equation for the vibro-acoustic wave

(i) *Vibroacoustography*: It has been established that a secondary acoustic radiation emanates from a region in a body experiencing a low-frequency radiation force owing to the mixing of two ultrasound waves in their intersecting focal volume, called here the ROI. This low-frequency acoustic wave is the VA wave [8]. The VA wave contains information on material properties of the ROI, namely, density, and shear modulus, and its size. The application for which VA became well known is imaging: images qualitatively representing shear-elastic information in the object could be created by measuring the variation in VA amplitude (or, in some cases, the phase) when the intersecting focal volume scans the object [9, 10]. There were also attempts to form quantitative images of elastic property variation by inverting the measured VA amplitude [11]. Vibroacoustography is the name given to this imaging modality where an 'image', either quantitative or a qualitative one which is only a representation of the elastic property, is arrived at from the measured amplitude. Since a quantitative image of the elastic stiffness would make vibroaoustography an important tool for diagnostic imaging, attempts to arrive at quantitative images through inversion are many (see, for example, [12]). Since an inversion of amplitude data has to take into account the effect of the intervening medium, acoustic inhomogeneities in the object, change in ultrasound force when different regions are scanned and a host of others, an accurate inversion is quite a complex job. Therefore, we have attempted an inversion based on a set of frequency measurements, the natural frequencies of vibration of ROI [7]. The natural frequencies are inferred from the VA signal from its spectral peaks. The locations of the spectral peaks of the VA signal are unaffected by propagation through inhomogeneous objects or an intervening medium. In order to justify equating a VA spectral peak to a natural frequency of vibration of the ROI, we need to establish a direct linear relationship between the secondary VA amplitude detected and the strength of vibration of ROI, or in other words, the shear-wave amplitude in the ROI and its neighbourhood. This we do in the following sub-section.

(ii) *Vibro-acoustic wave as a carrier of shear-displacement field*: An ultrasound wave in its passage through a body produces a sequence of compression and rarefaction in it, leading to the generation of low-frequency acoustic waves (the VA waves). Because of these, transverse shear waves could also be generated because of nonzero Poisson's ratio and/or material anisotropy, only as a secondary effect. With a view to preventing overheating of tissue, in the context of imaging, the applied ultrasound force is kept small. When the radiation force is small, amplitudes of both the waves are small and of the same order. (When the ultrasound intensity is large enough for tissue ablation, the shear-wave amplitude can be large which could then propagate to detectors far removed, in spite of absorption *en route*.) The weak shear waves have no chance of propagation beyond the neighbourhood of the ROI, owing to high level of attenuation by tissue. Our objective here is to show that the primary VA wave acts as a carrier to transmit shear-displacement information in the ROI and its immediate neighbourhood to regions where the detectors are located. Towards this, we first consider the linear momentum-balance equation for the ROI, which is

$$\rho\frac{D\dot{\mathbf{u}}}{Dt} = -\nabla p + \nabla \cdot \boldsymbol{\sigma}^{\text{dev}} \quad \mathbf{x} \in \Omega_{\text{ROI}} \tag{4.1}$$

Here, $\dot{\mathbf{u}}$ is the velocity of the material point $\mathbf{x}$, $\boldsymbol{\sigma}^{\text{dev}}$ is the deviatoric part of the Cauchy stress tensor, $p = p_0 + \tilde{p}_a$, which means the pressure is decomposed into an initial component p_0 and an ultrasound-generated part (owing to the acoustic wave) $\tilde{p}_a$. Moreover, $\Omega \supset \Omega_{\text{ROI}}$ is the domain of the ROI. Assuming that the initial pressure is spatially homogeneous, we write the gradient of p on the right-hand side of equation (4.1) as $\nabla\tilde{p}_a$. We note that $\tilde{p}_a$ which is originally produced in Ω_{ROI} picks up the time variations in the shear-dominated component of the displacement field and propagates to the entire object domain, Ω, as p_a obeying the wave equation:

$$\frac{1}{c^2}\ddot{p}_a = \nabla \cdot \nabla p_a \quad \mathbf{x} \in \Omega \tag{4.2}$$

Here, the acoustic velocity is given by $c^2 = G/\rho$, where G is the shear modulus of the material, and ρ its mass density satisfies the continuity equation, $\dot{\rho} + \nabla \cdot (\rho\dot{\mathbf{u}}) = 0$.

Since in the small deformation regime we are considering, material density is assumed a constant in space and time, the continuity equation does not play any role in the arguments that follow. We note that $\tilde{p}_a = p_a I_{\Omega_{\text{ROI}}}$, where $I_{\Omega_{\text{ROI}}}$ is the indicator function for Ω_{ROI}, and we can write p_a as $p_a = p_a^0 + p_a I_{\Omega_{\text{ROI}}}$. Then equation (4.2) can be re-arranged as

$$\frac{1}{c^2}\ddot{p}_a^0 - \nabla \cdot \nabla p_a = \nabla \cdot \nabla\tilde{p}_a - \frac{1}{c^2}\ddot{\tilde{p}}_a, \quad \mathbf{x} \in \Omega/\Omega_{\text{ROI}} \tag{4.3}$$

Note that the domain over which equation (4.3) is defined is restricted to $\Omega/\Omega_{\text{ROI}}$. This is justified because the linear momentum-balance equation, (4.1), is valid within Ω_{ROI} and the coupling effect of the wave propagation equation, (4.2), has negligible influence on $\mathbf{u}$. The source-term on the right-hand side of equation (4.3) contains information on the shape of the ROI (through the presence of I_Ω), shear-deformation signature and the density of the material within the ROI. These are then carried by p_a^0, the pressure wave, to everywhere in Ω, particularly to $\partial\Omega$, where we measure $p_a^0(\mathbf{r}, t)$. We measure the temporal spectrum of the measured pressure, which is seen to contain, along with density and shape of the ROI mentioned earlier, information on shear deformation within the ROI. This elucidation of the passage of shear-modulus information in a selected location through VA wave was first given in [7].

Since the VA signal is influenced by a host of other parameters than just the elastic properties of the object, in formulating a connection between the measured signal and the elastic properties (so that the elastic properties can be 'inverted' from the measurements) all the other influences have to be accounted for. Since the acoustic radiation force also enters into the inverse problem formulation, the variation of this force as the ultrasound focal volume scans the object, also needs to be accounted for. In the current literature, to the best of our knowledge, the force variation is not accounted for, and hence the accuracy of the recovery is suspect, to that extent. As carried out in the experiments reported in chapter 3, we shift our measurement to a set of frequencies present in the Fourier transform of the measured acoustic pressure, the location of the spectral peaks which are the resonant modes of the VROI. This presupposes (and is proven true in chapter 3) that the natural frequencies of the ROI correspond to the peaks in the power spectrum of $p_a^0(\mathbf{r}, t)$, $\mathbf{r} \in \partial\Omega$. This identification has been proven in [7] through simulations, and the underlying assumptions are: (1) the ROI is a freely vibrating object even though surrounded by the rest of the object, and (2) the medium between the ROI and the detector does not cause $p_a^0(\mathbf{r}, t)$ to suffer temporal modulation which would result in shift in the power spectrum. Owing to this, we gain computational advantage whilst verifying measured resonant frequencies in the acoustic signal, as they are the same as the natural frequencies of the ROI. We have employed ANSYS®, a commercial software, to compute the resonant modes of the ROI. In the sub-section below, we briefly explain the variational method employed in the commercial program to arrive at these frequencies.

4.2.2 Computation of natural frequencies

To recover shear modulus, we follow a strategy that is similar to that adopted in chapter 3, i.e. compute natural frequencies of the VROI, compare them with the measured ones, and minimize the mean-squared error by adjusting the elasticity parameters. In this sub-section, we briefly introduce how a variational approximation for an undamped, unforced mechanical system leading to an eigenvalue problem, can be used to solve for the natural frequencies of the system. Recovery of elastic properties, density, shape and size from natural frequency measurement has been studied earlier in a number of publications [13–15]. The principle employed states that a free body, i.e. one with stress-free boundary condition, which obeys the elastic wave equation, the Lagrangian, $L = \int_V (E_K - E_P)dV$ is stationary. Here, E_K and E_P are, respectively, the kinetic- and potential energy of the body, given by $E_K = \frac{1}{2}\rho\omega^2 u_i u_i$ and $E_P = \frac{1}{2}C_{ijkl}\partial_j u_i \partial_l u_k$ with $\mathbf{u} = (u_i, u_j, u_k)$ representing the displacement vector suffered. (Note that the Einstein convention of repeated indices is followed here.) Moreover, u_i, etc, have a harmonic time variation given by $e^{j\omega t}$ and C_{ijkl} is the elasticity tensor. We then expand u_i, etc, using the basis function set

$\{B_\lambda\}_{\lambda=1,\,2,\,\ldots N}$ defined over V and substitute in the condition, $\delta L = 0$. This leads to the symmetric eigenvalue problem, given by

$$\omega^2 \mathbf{\Gamma} \cdot \boldsymbol{\alpha} = \mathbf{M} \cdot \boldsymbol{\alpha} \tag{4.4}$$

Here, the mass matrix, $\mathbf{M}$, is symmetric, real and positive definite and the stiffness matrix, $\mathbf{\Gamma}$, is also symmetric and real but positive semi-definite. The elements of $\mathbf{M}$ and $\mathbf{\Gamma}$ are: $M_{i\lambda,\,i'\lambda'} = \delta_{i,\,i'} \int_V B_\lambda \rho B_{\lambda'} dV$ and $\Gamma_{i\lambda,\,i'\lambda'} = C_{ij,\,i'j'} \int_V B_{\lambda,\,j} B_{\lambda',\,j'} dV$. Again, $i, i', j, j' \in [1, 3]$, $\lambda' = 1, 2, \ldots N$ and $B_{\lambda,\,j}$ denotes the derivative of B_λ with respect to the jth spatial coordinate. The concatenated eigenvector, $\boldsymbol{\alpha}$ is obtained by arranging all the components, $\tilde{u}_{i,\lambda}$, of u_i.

Visscher *et al* [15] provided the lead to generate an easily computable forward model, applicable for any arbitrary-shaped object. They employed a basis function set made of monomial of the form $B_\lambda = x^{l(\lambda)} y^{m(\lambda)} z^{n(\lambda)}$. As explained in [7], this selection of basis function leads to closed form, analytic expressions, valid for objects of any shape, for the computation of the necessary matrix elements, preserving also the symmetry and positive-definiteness of $\mathbf{M}$.

We note that the software package ANSYS we employed here for the computation of natural frequencies and which also forms a part of the inversion iteration for recovering elastic properties from the experimentally measured natural frequencies, uses the same variational formulation and the monomial expansion for basis function as discussed above. The inputs to the ANSYS routine are: (i) geometry of the VROI delimited by the free boundary, (ii) material parameters such as ρ and C_{ijkl} (in this first section of the chapter, where we are dealing only with a linear, isotropic object, only Young's modulus and ρ) and (iii) boundary conditions appropriate for the ROI. For identification of the free boundary of the VROI, we follow the procedure already explained in chapter 3 (see sections 3.2.2–3.2.4). The free boundary is denoted by $\partial\Omega$ which separates the vibrating region from its non-vibrating complement.

With the help of this numerical tool for eigenvalue analysis, we now establish the connection between the dominant natural frequency of the VROI and the shear modulus of the material therein. The material parameters and the shape and size of the intersecting ultrasound focal volumes (the VROI) are so chosen as to match those that pertain to the experiments reported in section 4.3. In comparison to the experiments of chapter 3, the experimental setup here has a small modification. We use two separate ultrasound transducers, not a single confocal one; the object is an agarose tissue-mimicking phantom. The axes of the transducers meet at 90° and the nearly hyperboloidal focal regions meet at their thinnest waist. The forcing due to mixing of the ultrasound radiation is from the low-frequency acoustic wave, and this can be computed as $F(t) = F_0 \cos(2\pi \Delta f_0 t)$, where F_0 is independent of time but spatially varies within the ROI, and Δf_0 is the difference in frequency of operation the two transducers are adjusted for. (Computation of this acoustic radiation force is discussed in section 3.2.) The sinusoidal force distribution computed is then used in the momentum-balance equation (equation (3.1), which is repeated here for convenience)

$$\rho \ddot{\mathbf{u}} = \nabla . \, \sigma + F_0(\mathbf{x})\cos(2\pi \Delta f_0 \, t) I_{\Omega_{\mathrm{ROI}}} \tag{4.5}$$

Equation (4.5) is solved for $\mathbf{u}(t)$ one full cycle of sinusoidal forcing from which the amplitude of vibration is ascertained. The stress-free boundary, $\partial\Omega$, is obtained through identifying nodes, on the periphery of the ROI, which can be approximated to have zero amplitude of vibration.

We then repeated the above simulations for a homogeneous material mimicking the agarose slab used in the experiments reported in section 4.3. The low-frequency acoustic radiation force is assumed to be applied at the centre of the slab. Using an axis-symmetric analysis, with the help of ANSYS, we solved equation (4.5) for the displacement. The components of amplitude in three orthogonal planes passing through the centre of the object are shown in figure 4.1. It is clearly seen that the deformation owing to the local application of the radiation force is confined to a region around where the force is nonzero, which means the dynamics introduced is truly local. In other words, the forced vibration response is dominated by eigenvectors that best capture the ROI-localized deformed shape of the object.

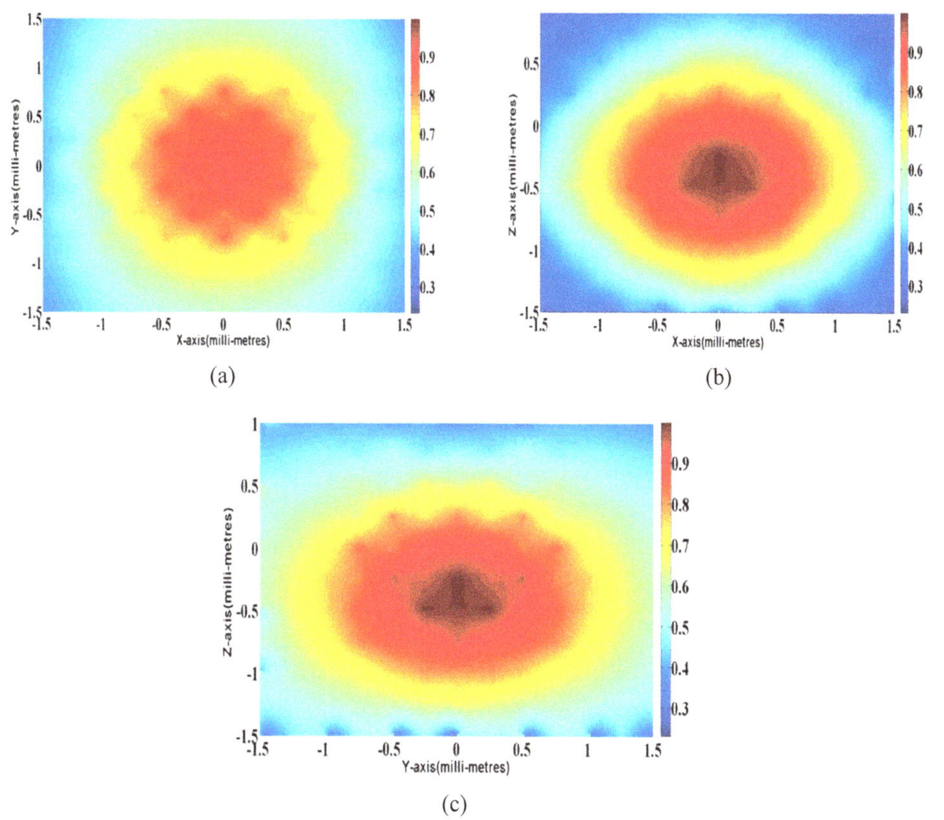

Figure 4.1. Axial cross-sections (a) X–Y, (b) X–Z, (c) Y–Z of the normalized displacement magnitude around the focal volume. It is seen that the displacements go to small values (blue) quickly. Reprinted with permission from [7]. © 2016 Institute of Physics and Engineering in Medicine.

Further, we have verified that the resonant modes of vibration of the VROI remained practically invariant when we solved the eigenvalue problem with the VROI enclosed by −3 dB, −4.5 dB, −6 dB and −7.5 dB surfaces of the central vibration amplitude. This proves that, for all practical purposes, the dynamics is confined to a small volume around the intersection of the ultrasound focal regions. This justifies the use of VA waves for measurement of local elastic properties within the object.

Having got the shape and size of the VROI, and with the knowledge of density and elasticity modulus of agarose, our next step is setting up of the eigenvalue equation (equation (4.4)) and solving it for $\{\omega^2\}$, the eigenvalues of the VROI. (In all our computations, we have allowed the −3dB surface of the vibration amplitude to determine the size and shape of the VROI.) With all this information in place, we used an appropriate modal analysis subroutine from ANSYS to do the inversion of equation (4.4). In the experiments reported in section 4.3, we have made agrose slabs of different shear moduli (all homogeneous) by varying the percentage by weight of agarose powder used in the solution. The recipe used to make agarose slabs is given in [7]. By increasing this percentage from 1 to 4, the storage modulus (G') of the samples varied from 20 kPa to 60 kPa; which is also verified by independent rheometer measurements. (The rheometer also gave us loss moduli, G''.) Within the validity of Kelvin–Voigt model, under which we can place both human tissue and tissue-mimicking agarose slab, when strain is assumed small, the storage modulus is numerically equal to shear modulus (G); also, with Poisson's ratio taken as 0.495, the Young's modulus is three times G [16]. Figures 4.2(a) and (b) give, respectively, the variation of G' and G'' against the amplitude of applied strain, corresponding to agarose samples with different weight percentage of stock powder.

From the eigenvalues obtained from equation (4.4), the variation of the first fundamental frequency with shear modulus is plotted, which is shown in figure 4.3. In the range of G we are interested, the variation is seen to be almost linear. A nonlinear fit using $\omega = a_0 + a_1 G + a_2 G^2$ is also shown in figure 4.3, and the coefficients a_0, a_1 and a_2 are found to be 130.36, 7.547 and −0.034, respectively, showing that the nonlinearity is mild. Also shown in the figure are the experimentally measured fundamental frequencies and the match is seen to be reasonably good.

4.3 Experiment to measure the natural frequencies

4.3.1 Experimental setup

For a schematic diagram of the experimental setup see figure 4.4. As is seen, the setup has three parts: (1) two ultrasound focusing transducers which together give a low-frequency acoustic force in a selected region, (2) the object to be interrogated mounted on translation stage and immersed in water to provide acoustic impedance matching, and (3) a low-frequency acoustic radiation detector, which in our case is an FBG sensor mounted on a cantilever.

The ultrasound transducers (Sonic Concepts, Washington) operate at a centre frequency of 1.1 MHz each delivering an average power up to 400 W and are driven

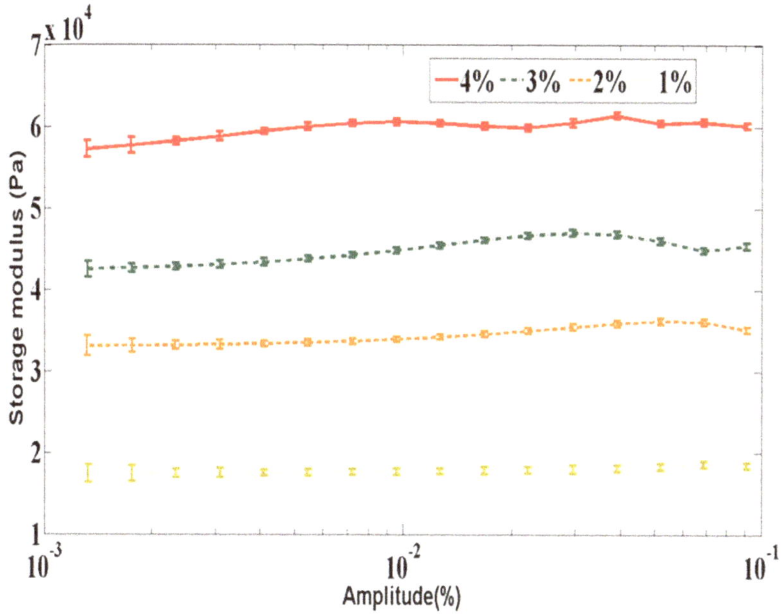

(a)

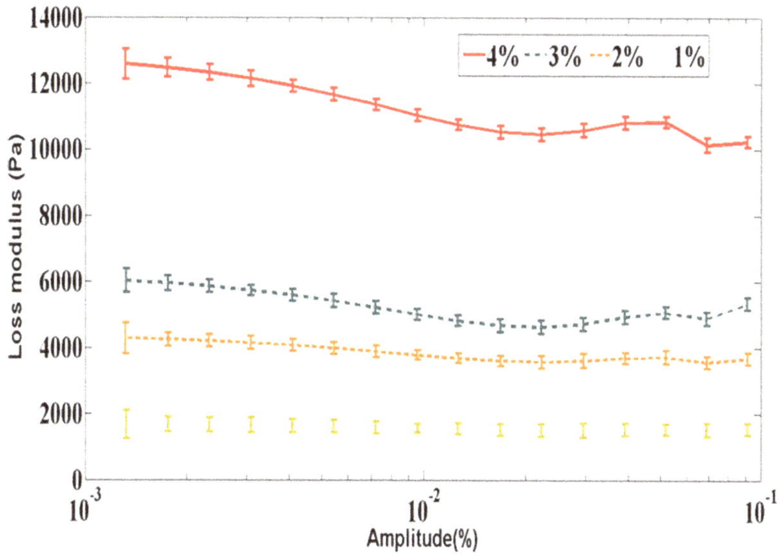

(b)

Figure 4.2. Experimentally obtained variations of storage modulus (G') (a) and loss modulus (G'') (b) with strain (from Anton Paar rheometer) for samples with 1%, 2%, 3%, and 4% (by weight) of agarose powder. For details of preparation of agarose samples, see [7]. Reprinted with permission from [7]. © 2016 Institute of Physics and Engineering in Medicine.

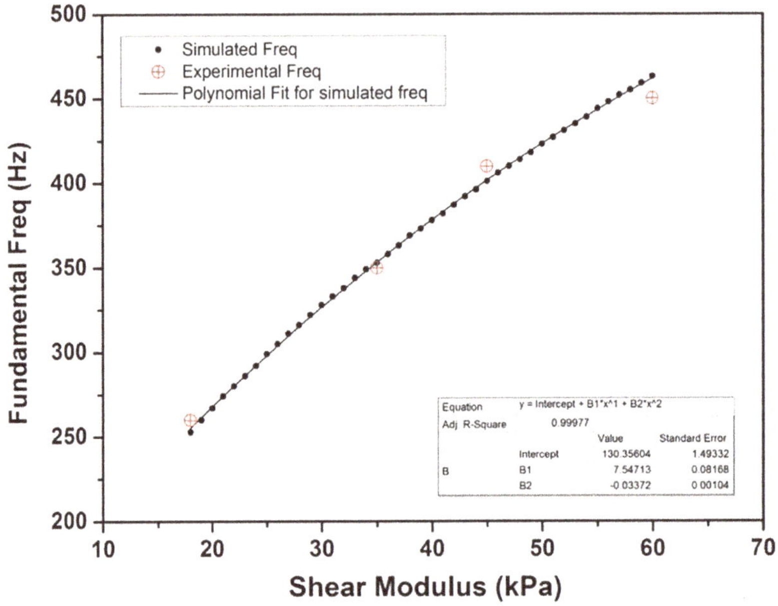

Figure 4.3. Variation of first modal frequency, both simulated and experimentally obtained, with shear modulus of the material in the ROI. Reprinted with permission from [7]. © 2016 Institute of Physics and Engineering in Medicine.

by a dual-channel function generator (Tektronix, AFG, 3022B) after suitable power amplification. Operating frequency of one of the transducers can be altered precisely to 1.1 MHz + $\Delta f_0/2$ and the other to 1.1 MHz – $\Delta f_0/2$, where Δf_0 can be varied from tens of Hz to 1 kHz. The transducers are identical focusing ones with f/no of 0.9781 which are precisely aligned so that their focal volumes intersect at an angle of 90° at their thinnest waist region. The volume of the intersecting region is found to be approximately 4.23 mm^3, which imposes the spatial resolution limit on the mechanical property images we may recover.

We prepared four slabs of agarose (dimensions 2 cm × 3 cm × 4 cm) with shear modulus 17 949.4 Pa, 35 082.7 Pa, 45 281.1 Pa and 59 983.9 Pa, verified through independent rheometer measurement. In our experiments, scanning with the intersecting ultrasound focal volume is organized by moving the slab with the help of a computer-controlled translation-cum-rotation stage (made by Holmarc, Cochin, India). The object and the transducers are immersed in water. For generating 2D cross-sectional image of shear modulus of, say, an excised organ, that plane in the organ is raster-scanned by the intersecting focal volume.

The VA wave emanating from the ROI and reaching the surface of the object is detected by an FBG sensor for pressure. The principle of the FBG, and its application in basic strain sensing are given in [17]. In brief, it is a thick phase grating written on the core of a photo-sensitive fibre through modulation of the refractive index, which selectively back-reflects a particular wavelength from an interrogating broad-band light input, provided that wavelength satisfies the Bragg

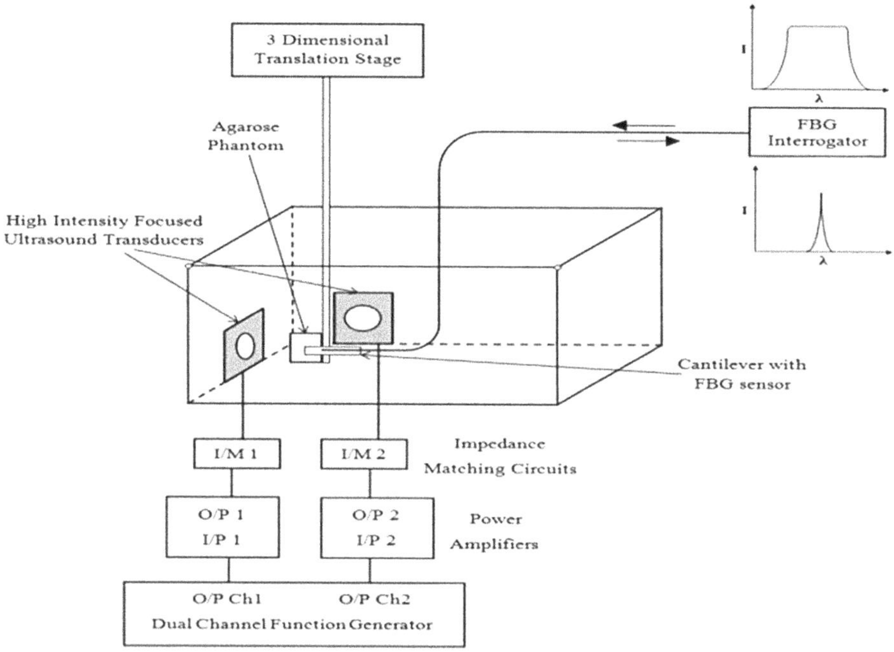

Figure 4.4. Schematic diagram of the experimental set-up. The two transducers are driven by signals from an ultra-stable, dual-channel function generator after power amplification. The transducers and object are mounted on translation stages. The object, as well as transducers, is immersed in a water bath for acoustic impedance matching. The acoustic wave is detected by an FBG sensor mounted on a cantilever. Reprinted with permission from [7]. © 2016 Institute of Physics and Engineering in Medicine.

condition. When the grating is subject to stress, the ensuing strain causes this wavelength to shift. This shift in wavelength is then calibrated against displacement (straightforwardly) or against strain. Displacement, in turn, can be related to other parameters such as pressure, temperature, etc, leading to FBG-based sensors for all such parameters. Here, by mounting the FBG on a cantilever with high sensitivity of movement towards incident pressure wave, we can calibrate and measure pressure variations from the resulting changes in the back-reflected Bragg-wavelength (λ_B). The electronic assembly which forms part of the FBG interrogator, detects this shift in wavelength to *pm* resolution, which is the readout from the sensor. For the pressure variations of our interest, we can assume a linear variation in the shift of λ_B; however, the amplitude of the VA wave is small and a highly sensitive transduction device, here a thin stainless steel cantilever, has to be designed so that FBG is subjected to a displacement large enough to produce a measurable λ_B. With this in place, the time varying VA wave produces a time-varying $\lambda_B(t)$. This is shown in figure 4.5(a). The power-spectrum of the signal in figure 4.5(a) is shown in figure 4.5(b), which contains, apart from the smaller peaks corresponding to spurious signals picked-up by the FBG, and also noise, a major peak at the ultrasound beat frequency, Δf_0. The average value of this peak amplitude is our measurement,

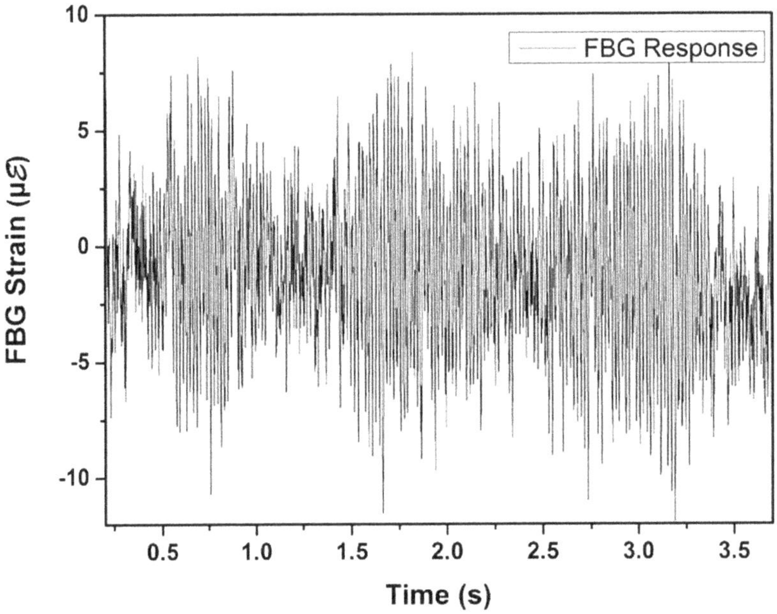

Figure 4.5a. Strain as read out by the sensor owing to the acoustic wave incident on the cantilever tip. Reprinted with permsssion from [7]. © 2016 Institute of Physics and Engineering in Medicine.

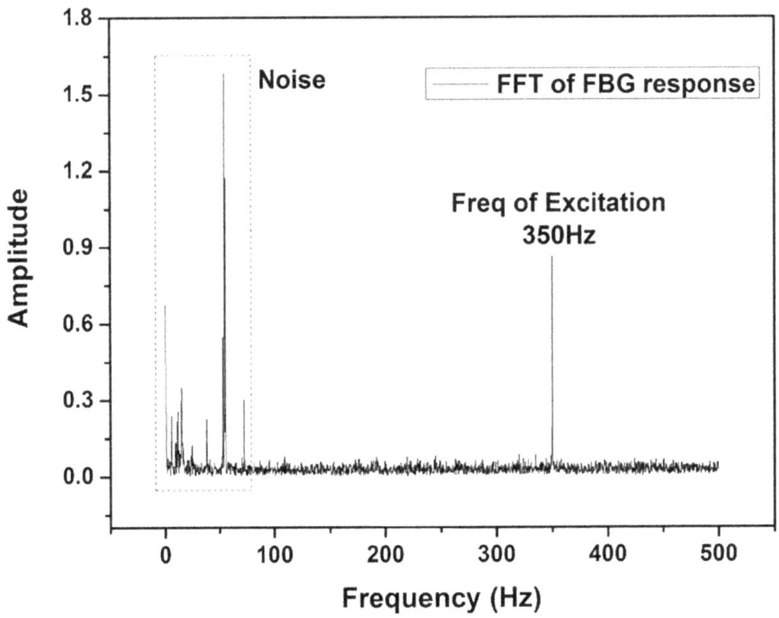

Figure 4.5b. Power spectral density of the data of figure 4.5(a). Amplitude at the difference frequency, here 350 Hz, is the measurement. Reprinted with permission from [7]. © 2016 Institute of Physics and Engineering in Medicine.

obtained by repeating the experiment for different locations of the detector around the object.

4.3.2 Determination of the first natural frequency of the region-of-interest

This part takes the same route followed in the first experiment described in chapter 3. The ultrasound beat frequency is swept from a low value to around 500 Hz, within which we suspect the first resonant frequency of the VROI is found, and at each frequency the amplitude of the peak at Δf_0 is measured. The limit of 500 Hz was owing to the FBG interrogator with a 1 kHz sampling rate used here. Plots of the peak amplitude versus Δf_0 are in figures 4.6(a)–(d). These plots correspond to objects with increasing percentage of agarose powder, and hence increasing G. The amplitude peak itself shows a (resonant) peak in each case which heralds the approach of ultrasound scan frequency to the first fundamental mode of the VROI. We see the frequency of the observed peak in figure 4.6 increases from 280 Hz to 410 Hz as G is increased from 18.0 kPa to 60.0 kPa. Since we could go only up to 500 Hz we missed the higher harmonics in the resonant spectrum of the VROI. However, since the higher

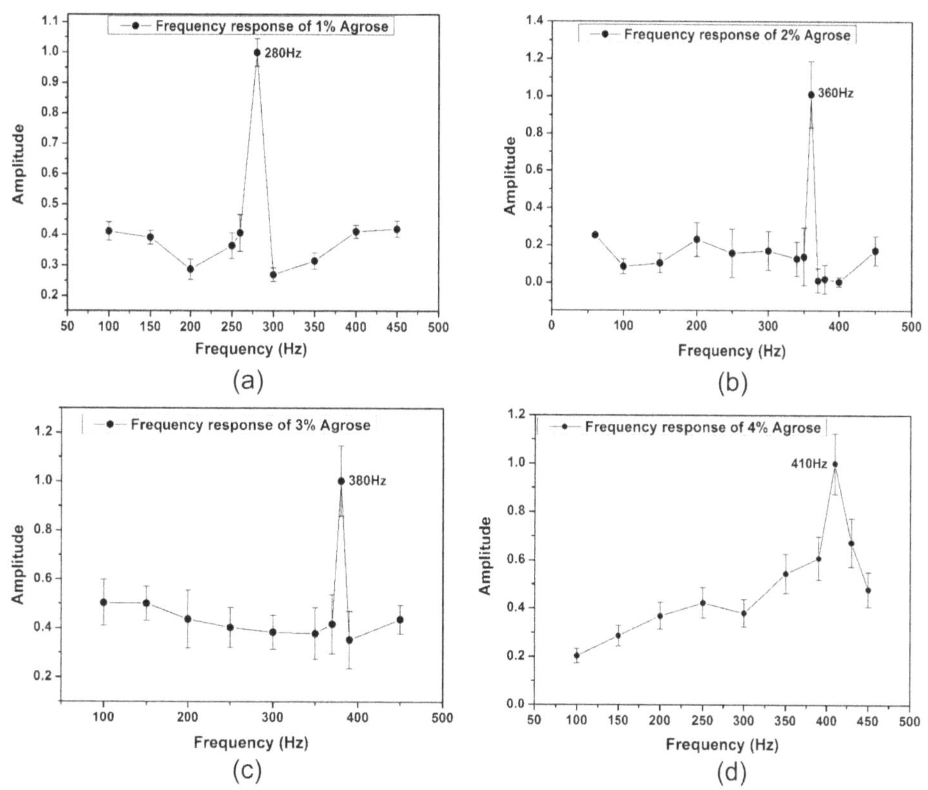

Figure 4.6. (a)–(d) Variation of the measured power-spectral amplitude at Δf_0 Hz (obtained from figure 4.5(b)) with Δf_0. The peak and its variation with object stiffness, as indicated by the percentage of agarose powder is evident. (a) 1% agarose, (b) 2% agarose, (c) 3% agarose and (d) 4% agarose. Reprinted with permission from [7]. © 2016 Institute of Physics and Engineering in Medicine.

harmonics are inherently weak and could not be observed through the experiments discussed in chapter 3, the chances of observing and measuring them are poor, even if we had an interrogator which could take us beyond 500 Hz. An amplification of the weak modes is required in order for them to be observed. This aspect is further discussed in section 4.4.1 of this chapter.

Table 4.1 gives the experimentally measured natural frequencies and their simulated counterparts, using $\partial\Omega$'s obtained as −3 dB, −4.5 dB, −6 dB and −7.5 dB contours of VA wave decay. It is observed that the match is reasonably good, especially when $\partial\Omega$ is the −3 dB contour.

4.3.3 Recovery of shear modulus from the natural frequency and discussion of results

Armed with experimentally measured data on resonant frequency of the VROI and a reliable route to compute them through an eigenvalue analysis, we are in a position to devise an inverse scheme to recover the elasticity parameters of the material of the ROI. For human organs, including changes brought about through progress of malignancy, tissue has to be modelled as at least orthotropic, needing a nine-component elasticity tensor to model 'stiffness'. Recovery of all these components needs a larger set of data: in our context a fuller set of the natural frequencies vibration of the VROI. We demonstrate the recovery of elastic tensor in the case of pork tissue in the second half of this chapter from a larger set of the measured natural frequencies, using an appropriate optimization scheme. However, presently, in the case of the agarose slab with isotropic elastic properties, we use the fundamental modal frequency measured to recover just one parameter, the average value of shear modulus within the ROI. For this simple recovery we use the method of bisection, already described in section 3.2.6.2 of chapter 3. For the four slabs, the recovered G values are given in table 4.2.

We have demonstrated here a quantitative VA procedure to map elastic property of tissue mimicking objects. This is made possible because the theoretical basis to verify the experimentally measured natural frequency of vibration is the creation of

Table 4.1. Comparison of the experimentally determined first mode of vibration with that from computer simulations using ANSYS. Reprinted with permission from [7]. © 2016 Institute of Physics and Engineering in Medicine.

Percentage of agarose	Experimentally measured first natural frequency (Hz)	Computed first natural frequency (Hz) with boundary from −3dB points	Computed first natural (Hz), with boundary from −4.5dB points	Computed first natural frequency (Hz), with boundary from −6dB points	Computed first natural frequency (Hz), with boundary from −7.5dB points
1	260	243	240	238	237
2	350	340	336	334	333
3	410	384	380	377	375
4	450	446	441	437	435

Table 4.2. Verification of shear modulus values obtained by the present method through standard rheometer measurements. Reprinted with permission from [7]. © 2016 Institute of Physics and Engineering in Medicine.

Percentage of agarose	Experimentally measured values of shear modulus from rheometer (Pa)	Reconstructed values of shear modulus (*G*) (Pa)
1%	17 949 (± 1093)	18 748.4
2%	35 082 (± 1514)	34 385.8
3%	45 281(± 985)	46 866.7
4%	59 983(± 1125)	56 734.3

local shear waves transported by the compressional VA wave. Since our measurement is a frequency, the noise which affects the VA wave does not affect our experimental measurement. In this context, it is also important to note that the magnitude of high frequency acoustic radiation force applied at a specific location depends on the acoustic absorption and scattering of the intervening object background. Therefore, the VA amplitude created and detected is also dependent on the location in the object interrogated by ultrasound forcing. If this dependence is not incorporated in the inversion algorithm, an amplitude-based recovery of elastic properties will be erroneous. In table 4.1, a slight discrepancy is observed between the computed and experimentally measured resonant modes. Since the computation of the frequency is dependent also on the volume and shape of the VROI, an error in the evaluation of the size and shape of the VROI affects the numerical accuracy of the computed resonant frequency.

4.4 Measurement of natural frequencies by ultrasound-assisted diffusing wave spectroscopy

Here, we replace the brute-force method presented earlier to measure the natural frequencies of the VROI (which was by scanning the ultrasound drive frequency and measuring the amplitude of the VA wave reaching the FBG detector) by a method involving diffusing wave spectroscopy (DWS).We have noted that the earlier method could not detect the higher-order modes which are characteristically weak. Measurement of these weak higher-order modes was also a challenge in the case of resonant ultrasound spectroscopy (in the application to characterize rock samples) as noted in [13]. In DWS, coherent photons are sent through the object, which, scattering many times from centres subjected to Brownian motion (in the background) and to Brownian motion as well as ultrasound-induced periodic oscillations (in the ROI) reach the detector. Here, the decay of temporal coherence suffered by the photons owing to random movement of the scatterers is measured and made use of to compute the mean-squared displacement (MSD) of these Brownian particles. This has been extended in [18], through studying the decay of modulation in the measured amplitude autocorrelation, to study the MSD pertaining only to the ROI. One of the important observations in [3, 18] is that the plateau of the MSD versus time curve acquires strong noise-like fluctuations when dynamics owing to ultrasound forcing is introduced within the object. These fluctuations are the result of mixing of two dynamics, one the temperature-induced Brownian and the other ultrasound-induced deterministic one, confined within the ROI. As mentioned in section 4.1 there is an amplification of the externally induced free vibration because of a cooperative (and resonant) transfer of energy from the background 'noisy' dynamics, making the otherwise weak higher order modes of the VROI standout against noise. This enhancement of the SNR of the sinusoidal signal with cooperative transfer of energy from noise, known as stochastic resonance (SR), has been extensively studied in the past [4, 5]. The SR is a resonant phenomenon because there is optimal noise strength when the SNR enhancement is maximized. In this particular case of multiple-scattering pathways of photons in the DWS probe, the scattering centres in each pathway can be considered to form a chain of Brownian particles. In the ROI, these particles form a chain of nonlinear oscillators, each particle attached to its neighbouring ones through springs and dashpots, i.e. with local linear coupling [6]. When there is an array of such oscillators, there is another parameter that influences the cooperative enhancement of SNR (apart from the noise strength) which is the coupling strength of the local oscillators to others in the chain. By tuning the coupling strength of oscillators one can further push the SNR enhancement, which is also a resonant phenomenon, for there is an optimal coupling strength when the SNR peaks for a local noise power [6]. This particular observation has been referred to as array-enhanced stochastic resonance (AESR). We propose that AESR is the reason behind the amplification of modes of vibration of the ROI. Work towards modelling the dynamics of a representative 'system particle' in the multiple-scattering photon path, which can simulate AESR, is still not complete, though presently we have some initial results which we present here. We will have a brief look at this again in sections 4.4.1 and 4.4.2, where the model we propose is further discussed.

The dynamics of a single system particle representing the ROI, with a set of so-called bath particles within the ROI affecting its dynamics, is modelled through the generalized Langevin equation (GLE) as is done in chapter 2, section 2.2.2. The set of bath particles, in our case some of them providing a chain of scattering centres in the diffusive photon path, in their interaction with the system particle, accounts for the visco-elastic environment of the medium. (Here, we identify each of the scattering centres (particle) in the zigzag photon path as a nonlinear oscillator with linear coupling with its neighbouring oscillators, as mentioned in the last paragraph.) In a recent publication from our group, the GLE was derived with a multiplicative noise term representing the effect of the nonlocal interactions on the system particle [3]. In [3] the ROI is modelled by a collection of harmonic oscillators, whose dynamics is represented by a 'system particle' (or, oscillator) with a predominant translational degree of freedom (DOF). The GLE is derived for this DOF with the effect of the remainder of oscillators (the so-called 'bath particles') captured by a multiplicative, or internal, noise term. This internal noise term which modifies the restoration term in the GLE captures the long-range, nonlocal interactions within the body owing to its material microstructure. These effects cannot be neglected when the space-time length scales of the forcing become comparable to the internal length scales of the material, which is true of both tissue and tissue-mimicking polymer such as the PVA or agarose. In addition, there is a term representing additive noise; the origin of this noise is temperature-driven fluctuation in the initial state of the particles. The associated dynamics is essentially Brownian, which helps us arrive at a correlation structure for the additive noise and also a fluctuation dissipation theorem (FDT), similar to the second FDT of Kubo [19].

The basic approach used in [3] is to capture micro-rotational, as well as translational, information pertaining to the particles by making use of a micropolar continuum setting to begin with. The micro-rotations flip from clockwise to counter-clockwise randomly because they are being acted upon by random forces from the background heat bath. The picture is similar to random flipping of spins discussed by Glauber in [20]. The transition probability of micro-rotations is also controlled by the instantaneous state of rotation of neighbouring particles which are connected to the system particle. This connection to the bath particles allows us to derive a correlation function between dynamics of particles and use it in the derivation of a modified FDT pertaining to the internal noise, as is done in [20]. Presently, our brief is only to offer an explanation to the observed amplification of natural frequencies of the VROI when measured using DWS. Towards this, we introduce a weak periodically time-varying acoustic force to influence the chain of scattering particles. By tuning the additive as well as multiplicative noise strengths, it is possible to make the acoustic forcing to take over the flipping of micro-rotation of particles with the frequency of flipping matching that of the externally applied periodic forcing. This is a resonant phenomenon and when the noise is tuned resonant transfer of energy takes place from noise to the 'weak signals' which makeup the spectrum of the VROI, and the weak higher harmonics become observable. For completeness sake, we very briefly introduce SR and AESR in the subsections that follow.

4.4.1 Stochastic resonance

Stochastic resonance occurs as the result of a cooperative interaction between a noise-induced and a signal-induced (deterministic) dynamics in a nonlinear system. The result is an enhancement of the output SNR when the 'intensity' or variance of input noise is tuned for a noticeable energy transfer from noise to signal. It has been observed that nonlinearity of the system plays a crucial role in producing this resonant energy transfer from noise to signal; for in a linear system the output SNR must equal the input SNR and an increase in input noise always decreased the output SNR. At the heart of the system showing forth SR there is a bistable (or, more generally, a multi-stable) dynamical device which can transition between the two stable states. The system is driven by a combination of a random force (i.e. noise) and a time periodic (in our case, sinusoidal) deterministic force, or signal. One can explain SR by considering an archetypal particle (a scattering centre) trapped in the symmetric double-well potential, a bistable device with two energy minima (see figure 4.7(a)). In the absence of periodic forcing, the heavily damped particle of mass m (in the host medium of viscous drag η) settles in one of the stable states with a small heatbath-induced transition probability of crossing over from one minimum to

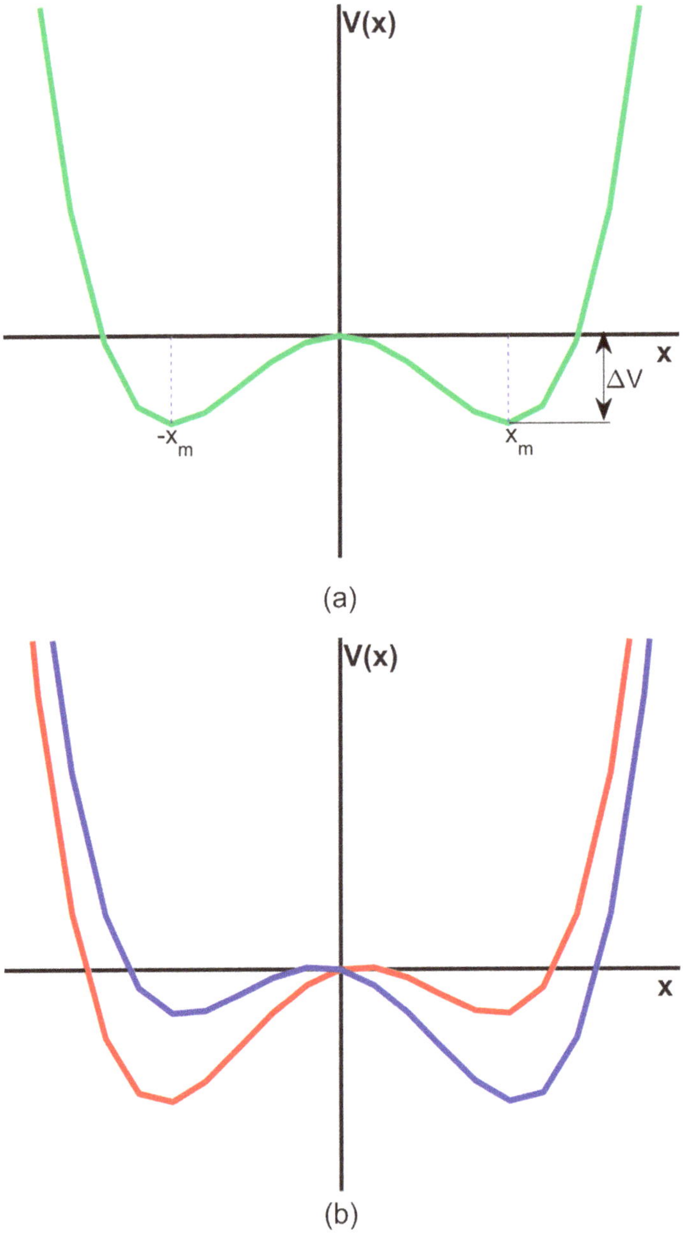

Figure 4.7. (a) Symmetric double-well potential, (b) the symmetry broken by the externally applied force.

the other. Since we are dealing with non-stationary signals in nonlinear systems (signals are rendered non-stationary because of the influence of the deterministic input), accurate theoretical description of stochastic resonance is quite complex. However, there have been many approaches to arrive at expressions for signal or SNR enhancement through proper modelling of the physics and signal processing aspects of SR. All work with approximations, the most common one with the restriction of a weak and low-frequency external signal, the so-called adiabatic approximation. Without fluctuations from noise the weak periodic forcing alone will be insufficient to force the particle to surmount the potential barrier to reach the other minimum. With noise alone, when the intensity increases there will be increased but random cross-over events. This cross-over rate is given by the well-known Kramer's rate [21]:

$$r_k = \frac{1}{\sqrt{2\pi}} \exp\left(-\frac{\Delta V}{D}\right) \tag{4.6}$$

Here, D is the variance of noise, a measure of noise intensity. When an external periodic signal (in our examples, sinusoidal), usually weak, is added to noise to drive the bistable system, the temporal variation in the sinusoidal force upsets the symmetry in the quartic potential, as shown in figure 4.7(b), owing to the periodic biasing towards one of the wells. The biasing follows the same frequency as the input sinusoid and the degree of bias can be controlled by D. It is here a form of coherence or 'cooperation' can be achieved between noise and signal, in that there is an optimal value of D (denoted by D_{opt}) for which the SNR of the output is maximised. At this noise intensity there is a resonant transfer of energy from noise to signal, and the transition of particle from the two minima follows the sinusoidal frequency, 'coherently' assisted by noise. Should we increase noise beyond D_{opt}, the cooperation ceases and noise takes over, indicated by the randomness of the flips.

This increase in SNR vis-a-vis the input noise strength has been explained through a number of approaches [4, 5, 22]. Many have used a rate-equation-based approach, wherein Kramer's hopping rate, which depends on noise intensity and modified by the external slow periodic input, was tuned to theoretically simulate the observed resonant maximization of SNR. The bistable system at the heart was either modelled a simple discrete two-state device or a continuously varying quartic potential well. In one set of approaches, the nonlinear system has been linearized using a perturbation approach which assumes a small-signal approximation. For the validity of this approximation, A, the amplitude of periodic input, should be kept small. The output SNR at the fundamental frequency of the periodic input has been arrived at as

$$SNR_{\text{out}} = \left(\frac{Ax_m}{D}\right)^2 \exp\left(-\frac{\Delta V}{D}\right) \tag{4.7}$$

However, there are instances of SR, which are outside the purview of linear response theory for modelling [23]. Without the use of small signal approximation, and the need for linearization, in [24] the output SNR is arrived at for a system with Gaussian noise. In yet another approach, the nonlinear system (stochastic dynamical system) is assumed to have Markovian property and the associated Fokker–Planck equation is solved for second order statistics of the output, which are useful for not only studying SR, but also for arriving at an FDT obeyed by the system [25]. The above advantages notwithstanding, the approach using the Fokker–Planck equation still requires certain approximations [25].

Further work needs to be done in order to simulate SR-based amplification and detection of the weak higher modes of the VROI. Experiments with a coherent light probe have conclusively proven (see section 4.4.3) that DWS-based data contains information on the amplitude and frequency of these higher harmonics of vibration. To complete this work, we intend to follow the stochastic signal processing approach presented in [22]. Here, we briefly touch upon the route taken in [22] to arrive at an expression for the output SNR. A periodic deterministic signal, $x(t)$ with period T_{in}, along with a stationary random noise, $\nu(t)$, is input to a nonlinear time-invariant system. The output, $y(t)$, is a cyclostationary random signal with period T_{in}, which is decomposed as $y(t) = \tilde{y}(t) + E\{y(t)\}$, where $\tilde{y}(t)$ is the fluctuating part of $y(t)$ and $E\{y(t)\}$ is its non-stationary mean. (Here, E is the expectation operator.) Denoting $E\{y(t)\}$ by $y_{av}(t)$, this non-stationary mean being a deterministic function of time, admits a Fourier expansion,

$$Y_{\text{av}}\left(\frac{n}{T_{\text{in}}}\right) = \frac{1}{T_{\text{in}}}\int_0^{T_{\text{in}}} y_{\text{av}}(t)\exp\left(-j2\pi\frac{n}{T_{\text{in}}}t\right)dt \tag{4.8}$$

The second-order average of $y(t)$, namely its autocorrelation, given by

$$E\{y(t)y(t+\tau)\} = E\{\tilde{y}(t)\tilde{y}(t+\tau)\} + E\{y(t)\}E\{y(t+\tau)\} \tag{4.9}$$

is a deterministic function of two variables t and τ. In t, this autocorrelation is periodic with period T_{in}. It is possible to average out the dependence on t by integrating over one time period T_{in} which would help us get the autocorrelation function $R_{yy}(\tau) = \frac{1}{T_{\text{in}}}\int_0^{T_{\text{in}}} E\{y(t)y(t+\tau)\}dt$. Using equation (4.9) we can express $R_{yy}(\tau)$ as $R_{yy}(\tau) = C_{yy}(\tau) + \frac{1}{T_{\text{in}}}\int_0^{T_{\text{in}}} E\{y(t)\}E\{y(t+\tau)\}dt$. Here, $C_{yy}(\tau)$, the stationary auto-covariance function of $y(t)$ is given by $C_{yy}(\tau) = \frac{1}{T_{\text{in}}}\int_0^{T_{\text{in}}} E\{\tilde{y}(t)\tilde{y}(t+\tau)\}dt$. With these functions in place, we are in a position to define the power spectral density (PSD) of $y(t)$, which is the Fourier transform of $R_{yy}(\tau)$, i.e.

$$\begin{aligned}\mathbf{R}_{yy}(\omega) &= \int_{-\infty}^{\infty} R_{yy}(\tau)\exp(-j\omega\tau)d\tau \\ &= F.\,T\{C_{yy}(\tau)\} + \sum_{n=-\infty}^{\infty} Y_{av}(\omega_n)Y_{av}^{*}(\omega_n)\delta(\omega-\omega_n)\end{aligned} \tag{4.10}$$

Here, $\omega_n = 2\pi\frac{n}{T_{\text{in}}}$ and F . T denotes Fourier transform operation. Following [22], we can decompose $C_{yy}(\tau)$ into $C_{yy}(\tau) = C_{yy}(0)h(\tau)$. Here, $C_{yy}(0)$, also denoted by $\text{var}[y(t)]_{\text{av}}$, is the stationary variance of $y(t)$ and $h(\tau)$ gives the normalized shape of $C_{yy}(\tau)$. (Stationary variance is the time-averaged version of the non-stationary variance of $y(t)$ defined as $\mathrm{var}[y(t)] = E\{\tilde{y}(t)\tilde{y}(t)\}$.) The function $h(\tau)$ admits a Fourier transform which is denoted by $H(\omega)$. With this the power spectral density of equation (4.10) can be expressed as

$$\mathbf{R}_{yy}(\omega) = \mathrm{var}[\,y(t)]_{\mathrm{av}}H(\omega) + \sum_{n=-\infty}^{\infty} Y_{\mathrm{av}}(\omega_n)Y^*_{\mathrm{av}}(\omega_n)\delta(\omega - \omega_n) \tag{4.11}$$

With equation (4.11) we are in a position to define the SNR corresponding to each individual Fourier component in the output, ω_n.

$$\mathrm{SNR}_y(\omega_n) = \frac{Y_{\mathrm{av}}(\omega_n)Y^*_{\mathrm{av}}(\omega_n)}{\mathrm{var}[\,y(t)]_{\mathrm{av}}H(\omega_n)\Delta B_n}. \tag{4.12}$$

Here, ΔB_n is the non-zero, small bandwidth around ω_n (the same as equation (11) in [22]). Evaluation of the SNR of equation (4.12) requires finding $y_{\mathrm{av}}(t)$ and $C_{yy}(\tau)$, the non-stationary mean and autocovariance of the output, respectively.

In the present context, we have a representative system particle driven by a background heat bath and the sinusoidal forcing from the ultrasound radiation. Driven by these two forces, one random and the other deterministic and sinusoidal, the output measured by DWS is an average quantity, the mean-squared displacement (MSD) of the system particle or its prominent component as a function of the time delay, τ, after the start of observation. The nonlinear system driven by the two forces is the GLE which outputs $u(\tau)$, time variation of the prominent component of displacement of the system particle. In particular, ours is a steady-state output as τ tends to larger values. The GLE was discussed in chapter 2 (equation (2.1)) which is reproduced here for convenience.

$$m\ddot{u} - ku + k'u^3 + \int_0^t \eta(s - t)\dot{u}(s)ds = F(t) + \xi(t) + \int_0^t W_s^t(s - t)u(s)ds \tag{4.13}$$

Here, the integral term on the left-hand side is the history-dependent viscous-drag term, The random forcing owing to heat bath, represented by $\xi(t)$, is the additive noise assumed to be white Gaussian. $F(t)$ is the low-frequency sinusoidal force from the ultrasound. The integral term on the right-hand side is the internal (or multiplicative) noise term representing overall effect of 'nonlocal interactions' of the bath particles (see also appendix A). The restoring force, $-ku + k'u^3$, is from the symmetric double-well potential from the acoustic trap provided by the interfering acoustic waves at the intersection of the focal regions of two synchronously run ultrasound transducers. (Experimental implementation of the acoustic trap in the present context is in section 4.4.3; a similar device was used in chapter 3 as well.) The double-well potential is shown in figure 4.7, a 1D version of it is given by $V(u) = \frac{1}{4}k'u^4 - \frac{1}{2}ku^2,\ k', k > 0$.

Solution of equation (4.13) for $u(t)$ helps us compute $\delta u(t) = u(t) - u(0)$ and the MSD $E[\delta u^2(t)]$. A typical plot of the growth of MSD with time delay is shown in figure 4.8. We would like to draw the reader's attention to steady-state fluctuations seen on the plateau whose power spectrum reveals that the SNR improvement

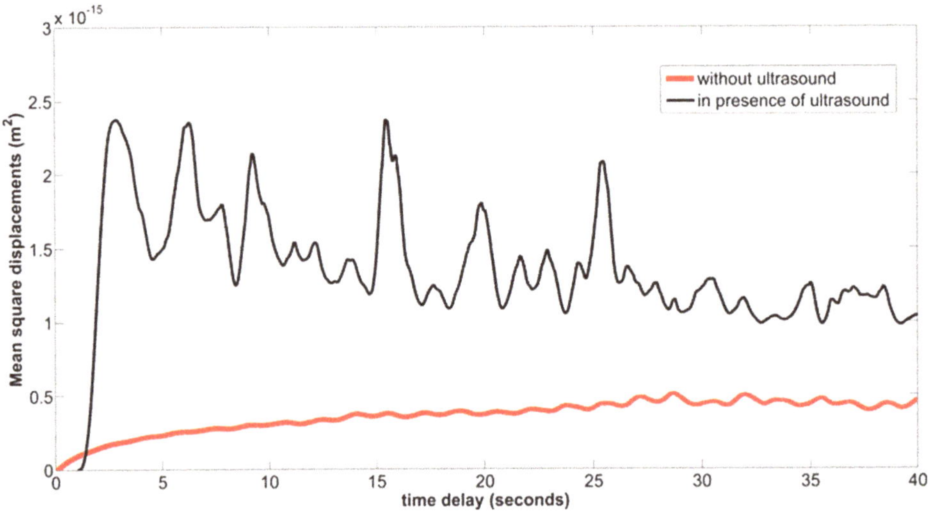

Figure 4.8. MSD versus t plot without ultrasound and with only additive noise in the model (red curve). The plateau fluctuations are seen to be weak. The black curve corresponds to when ultrasound is on and the model had a multiplicative noise, *not tuned* for resonance. The fluctuations in the plateau are relatively strong. Reprinted with permission from [32]. (2017) Optical Society of America.

(or, amplification), without the ultrasound on, is very little (red curve). The higher-order modes are simply not seen. It is clear that the introduction of a dual-beam ultrasound trap provides the bistable potential well, an essential component in the system that shows SR. Even with the ultrasound on, our simulation results show that tuning of the additive noise strength did not further improve the SNR of the VROI modes present in the plateau fluctuations. We have not yet completed this study, and therefore we do not have the final answer to explain the absence of SR with additive noise tuning. We conjecture that the array formed by light paths interconnecting scattering particles could not in unison show resonant oscillations transferring energy from additive noise alone. (There are two sources of additive noise in the ROI: one is the background temperature induced Brownian motion of the particles and the other is the dynamics due to high frequency pressure wave from the interfering ultrasound beams. As pointed out in [28] this high-frequency vibration provides an additive noise whose intensity is easily tunable, which can be used to demonstrate experimentally, the possibility of SR by additive noise manipulation.) However, we have demonstrated array-enhanced SR by tuning multiplicative noise strength through simulation; see figure 4.9 showing SNR variation and resonance with adjustment of σ_m^2. This demonstrates that, in the present context, to improve the SNR amplification to noticeable levels, we need to incorporate the nonlocal interactions between the system particle and the set of bath particles in the ROI. In other words, as indicated earlier in section 4.4.1, we need to consider arrays of scattering centres forming zigzag photon paths and coherent transfer of energy from fluctuation of the array to weak sinusoidal dynamics superimposed on the chain of scattering centres within the VROI. The possibility of optimizing this transfer

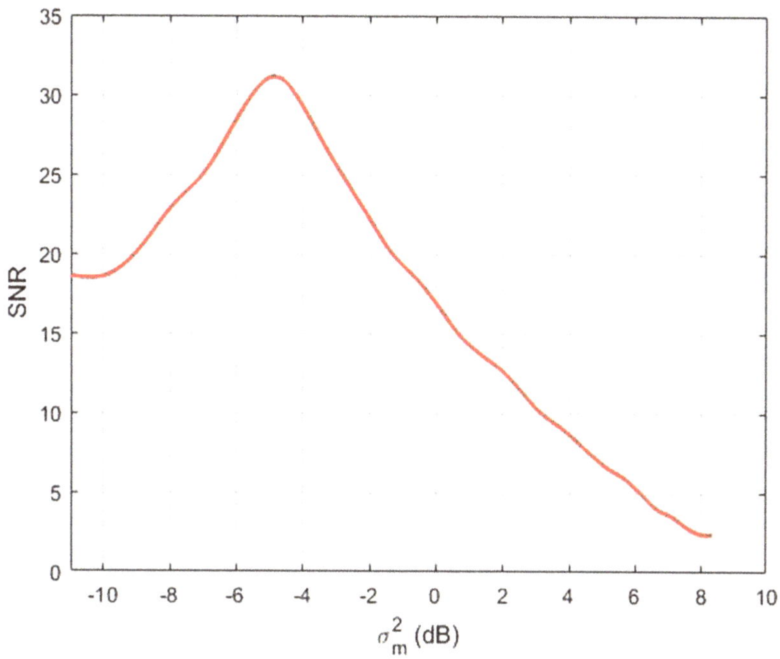

Figure 4.9. SNR of the first natural frequency of the VROI plotted against the intensity of multiplicative noise (σ_m^2) for a certain additive noise intensity (σ_a^2). (This additive noise intensity did not significantly shift or enhance the resonance observed with σ_m^2 tuning.) When σ_m^2 is increased beyond the resonance point the cooperation ceased and noise took over and the SNR began to fall.

demonstrated here through tuning the strength of interaction (i.e. multiplicative noise intensity) is array-enhanced stochastic resonance (AESR) mentioned in section 4.4.1. In the following section we discuss AESR within the background of ultrasound-assisted DWS and explain its role in making the natural modes of the VROI detectable. Even without optimizing, with the introduction of multiplicative noise of some intensity, there is an enhancement of plateau fluctuations (black curve in figure 4.8).

4.4.2 Array-enhanced stochastic resonance in the context of ultrasound-assisted diffusing-wave spectroscopy

The SR discussed in the previous section brings to light the constructive role noise can play in enhancing weak signals, in particular weak periodic signals. This cooperative role has to be forced by tuning average length of stay of particles in potential minima under the influence of stochastic forcing, to match half the period of the weak periodic signal. The resonance to maximize SNR enhancement by a proper selection of additive noise intensity is not yet demonstrated in the present context; further simulations to fine tune the role of various other parameters involved are needed to clearly see the SNR amplification and resonance. However, in the context of ultrasound-assisted DWS, even if there is an enhancement in the SNR with transfer of energy from additive noise alone, it is not noticeable, evident from the relatively weak fluctuations in the plateau of the MSD curve in figure 4.8 (red curve). As mentioned in section 4.4.1, when there is an array of nonlinear oscillators with linear coupling to neighbouring ones, oscillations are set up in the interconnected array as a collective response of the entire array to the combination of noise and signal. Tuning of the noise brings forth a resonance when the signal amplification is maximized. This array-enhanced SR (AESR) has been employed to explain signal transmission and detection in a number of areas [26] including ion channels, coupled network of certain neurons, arrays of Josephson junctions and multilevel threshold systems.

Noisy fluctuations observed in the plateau of MSD versus time delay curve are simulated by solving the GLE with a multiplicative (internal) noise in [3]. The structure and intensity of this internal noise were arrived at through (1) explicit derivation of Hamiltonians for the system particle and the set of bath particles in the background and (2) using the bath Hamiltonian to account for nonlocal interaction of the bath particles with the system particle. As pursued in [27], such interaction can also be modelled through terms like $k_i(x - x_i)$ for each bath particle at location x_i which is equivalent to projection of the higher dimensional Hamiltonian to the (x, p) subspace. In [3] the dynamics of the ROI is looked at using the background of micropolar continuum formulation, which takes into account not only the standard translation-based kinematics of the object, but also the nonstandard one based on micro-rotation. It is postulated (and verified) that nonlocal interactions in the object can be accounted for and quantified through the micro-rotations. As mentioned earlier (section 4.4.1) there is an inherent randomness in including micro-rotations as well in the dynamics owing to the fact that the translational and rotational strain operators do not commute, paving the way for an uncertainty relation. The internal noise in the GLE arises from this uncertainty relation. The main contribution of [3] is the characterization of this noise, its distribution (Gaussian) and strength which modulates the stiffness constant k. It was also demonstrated that experimentally observed fluctuations in the MSD plateau can be reproduced through tuning the strength of this internal noise appropriately.

In fact we can do far better than tweaking the noise strength to match the experiment, as demonstrated here. We can tune the internal noise strength to maximize the SNR of the modes (weak as they are) present in the VROI, and this signal amplification is attributed to AESR. We simulate AESR by solving the GLE and tuning the internal noise (as derived in [3]) and prove that strength of the internal noise is a parameter with which we can control and maximize SNR enhancement. Simulation results are shown in figures 4.9–4.12 which indeed reveal resonance and maximization of SNR.

We have been explaining nonlocal connections between particles by invoking micropolar theory, which assigns to particles two DOFs, namely, translational and rotational. Concentrating on microrotations suffered by the particles, the state of the system can be described by the set $\{\theta_i\}$, $i = 1, \ldots N$, where θ_i is the microrotation suffered by the ith particle measured on the 2D plane onto which it is projected and N is the total number of scattering particles in the ROI intercepted by light paths.

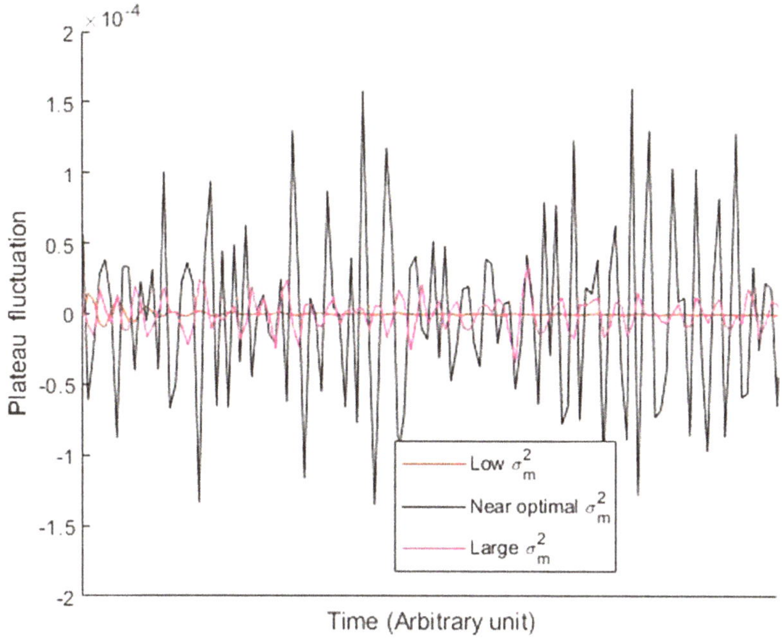

Figure 4.10. Plots of MSD versus lag time obtained with tuned σ_m^2 as well as those on either side of it. It is clearly seen that the fluctuations increase in intensity at resonance. This also can be inferred from the power spectra shown in figure 4.11.

The parameter θ_i of the ith particle in a linear visco-elastic continuum obeys a rotational GLE (similar to equation (4.13) for translation):

$$I\ddot{\theta}_i - \tilde{k}\theta_i + \tilde{k}'\theta_i^3 + \int_0^t \eta(s-t)\dot{\theta}_i(s)ds = F(t) + \xi(t) + \int_0^t W_s^t(s-t)\theta_i(s)ds \quad (4.14)$$

This is a modified version of equation (1) of [29] describing the evolution of the state θ_i. The linear connection terms representing nonlocal interactions are accounted for by the internal noise term, which is the integral term on the right-hand side. Moreover, I is the moment of inertia of the particle and $V_x(\theta_i) = -\frac{1}{2}\tilde{k}\theta_i^2 + \frac{1}{4}\tilde{k}'\theta_i^4$ is the double-well potential created by the interfering ultrasound beams in the ROI. Particles within the trap settle at one or the other of the minima, say $\theta_{i,m}$ or $-\theta_{i,m}$ with equal probability, and therefore the mean, $\langle\theta_i\rangle = 0$. It has been proven that multiplicative noise with intensity in a specified range can break this symmetry and create order so that $\langle\theta_i\rangle$ becomes nonzero, either positive or negative [29, 30]. The reason for this spatial ordering resulting in break of symmetry is that short- and long-range interaction of the particles can be captured by a multiplicative noise; and the coupling of particles favours neighbouring particles to stay in the same probability density maximum, i.e. in the *same minimum* of stochastic potential. In addition, the strength of multiplicative noise can introduce, or manipulate an already existing bistable potential for the advantage of SNR enhancement [30]. With the available co-existing bistable states (i.e. $\theta_{i,m}$ and $-\theta_{i,m}$) and a bistable

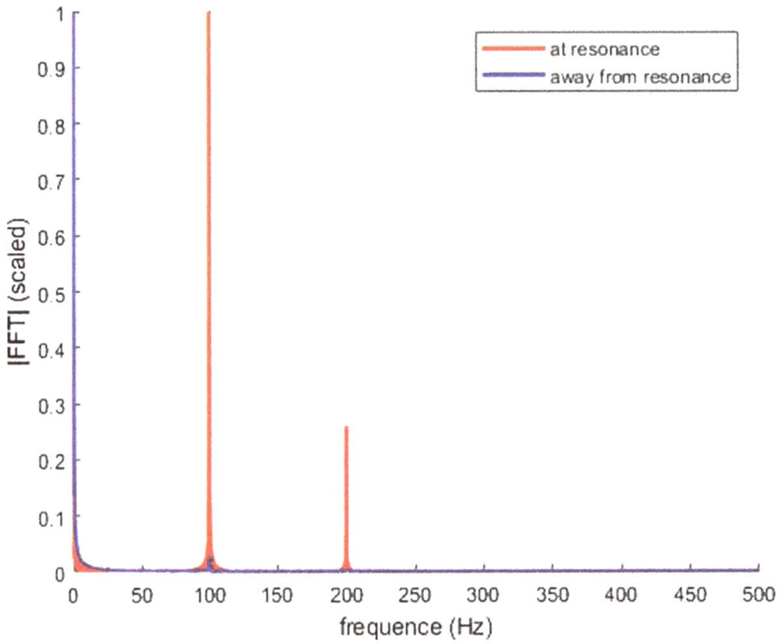

Figure 4.11. The power spectra of the plateau fluctuations shown in figure 4.10. It is seen that with acoustic force at 100 Hz the output of the nonlinear system at this fundamental frequency and the harmonics is greatly enhanced at resonance. This should be true also when the driving frequency is scanned to match any of the modal frequencies of the VROI. This extra computation is not done in the present work.

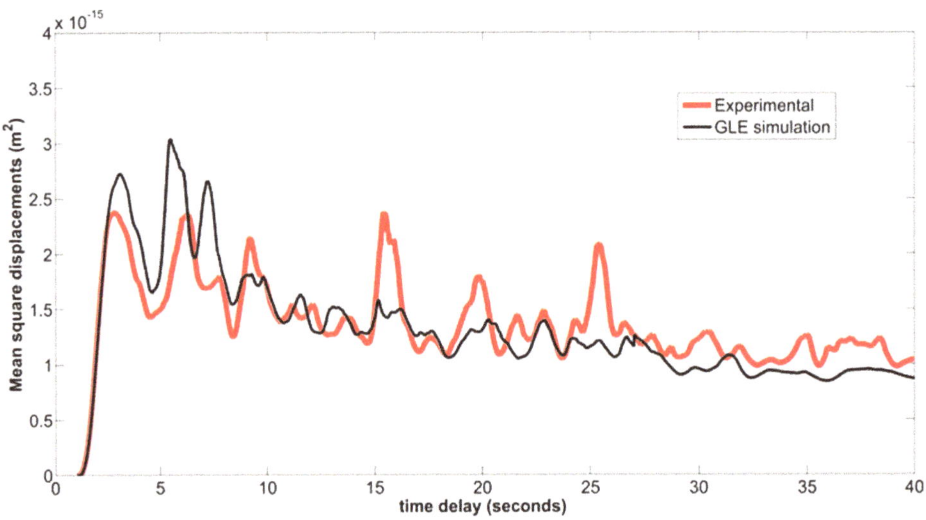

Figure 4.12. Plot of MSD versus delay time obtained from pork-fat slab compared with that from solving the GLE with a suitably tuned multiplicative noise. The match in the plateau fluctuations is seen to be reasonably good. This shows that the array interaction can be captured by a proper selection of the internal, multiplicative noise. Reprinted with permission from [32]. (2017) Optical Society of America.

potential, it is possible for the nonlinear system described by equation (4.14), and driven by a weak sinusoidal forcing and additive noise, to synchronize an output which is periodic with the same period as the input. The periodic output contains an amplified version of the input sinusoid. Moreover, matching the characteristic timescales of the coupled nonlinear oscillators results in the synchronous phenomenon of spatial ordering. These two effects simultaneously observed in the same nonlinear system, wherein favouring of an observation can be tuned to a maximum is referred to as doubly stochastic resonance (DSR) [31]. The name DSR is chosen because, in the first place there is a break of symmetry in state of the system through a cooperative effort of the multiplicative noise, nonlinearity and bistable potential, from the two equally probable states into which the system has settled down. In the second, there is a noise-assisted periodic transition between these two states tuned to an externally introduced weak periodic signal. It is beyond the scope of this monograph to provide a detailed discussion on these noise-assisted cooperative effects in nonlinear systems. The interested reader can refer to the already pointed out literature for further details. Results of our simulations demonstrating AESR in ultrasound-assisted DWS are shown in the figures below. Figure 4.9 is a plot of SNR at the first fundamental frequency of the VROI versus σ_m^2 showing forth clearly a peak in the SNR for a particular optimal value of σ_m^2. We have computed MSD using this optimal σ_m^2 and on either side of this optimal value. The resulting fluctuations seen in the plateau of MSD curve are plotted in figure 4.10. It is clearly seen that for $\sigma_m^2 = (\sigma_m^2)_{opt}$ the intensity of fluctuations is relatively high and easily noticeable. The power spectra of the MSD fluctuations are shown in figure 4.11 showing enhancement of the fundamental frequency which is the external acoustic drive frequency (100 Hz) and its harmonics at resonance. If we had scanned the external frequency to match any of the modal frequencies of the VROI the amplification of these also would have been evident. In figure 4.12 we show the MSD plot obtained from experiments on a pork slab with the one simulated from the GLE with σ_m^2 tuned so that the fluctuations in the plateau matched within error limits imposed by uncertainty in ascertaining scattering centre size and shape. Since the match can be considered acceptable, it is concluded that multiplicative noise representing inter-particle interaction has a large say in the way the deterministic and stochastic dynamics mix in the plateau region. This mixing is primarily because of the nonlinearity of the system, but influenced greatly by nonlocal interaction of the particles, here modelled by a multiplicative noise.

4.4.3 Experiments

The experiments are fully described in a paper from our group [32] and therefore some of the details are omitted owing to lack of space. These are done using either tissue-mimicking phantom or pork tissue. The phantom used is made of agar, samples of 65 mm × 17 mm × 18 mm slabs of increasing stiffness. Stiffness is manipulated by adding increasing percentage by weight of agar powder in the stock solution and turbidity is created by mixing intralipid. Transport mean-free-path (l^*) is maintained at approximately 1.4 mm (for details, see [32].) Altogether three samples were used, whose storage- and loss moduli spectra (up to 100 Hz) were measured using a rheometer (Discovery, HR3 from TA Instruments) measurements. The second set of slabs of the same dimension as agar are cut from freshly harvested pork fat preserved in a phosphate-buffered saline solution at 27 ± 0.5 °C.

4.4.3.1 Diffusing-wave spectroscopy with ultrasound tagging

The experimental setup used here is quite similar to the one described in section 3.2.6.2. Figure 4.13 shows the present setup, in which, instead of one confocal ultrasound transducer we used two single-frequency ones, driven synchronously by a dual-channel function generator. Light from a laser source illuminates the object under study. Photons reaching the opposite side are detected by a photon-counting PMT (Hamamatsu, H7360-03). The weak current signal from the PMT is converted to a voltage signal and sent to a signal conditioning system and amplifier. The amplified signal is input to a digital autocorrelator (DAC, Flex 021d from Correlator.com, with the least sample time of 12.5 ns) which computes a time autocorrelation of the signal. The two synchronously driven focusing ultrasound transducers, forming the acoustic part of the setup, (V394-SU-F50MM-PTF, Olympus NDT NE, Inc) are mounted on $x - y - z - \theta$ translation stages. The micro-positioners help us align the transducers so that their focal regions meet at 90° angle at its thinnest 'waist' region within the object which is either the agar or pork-

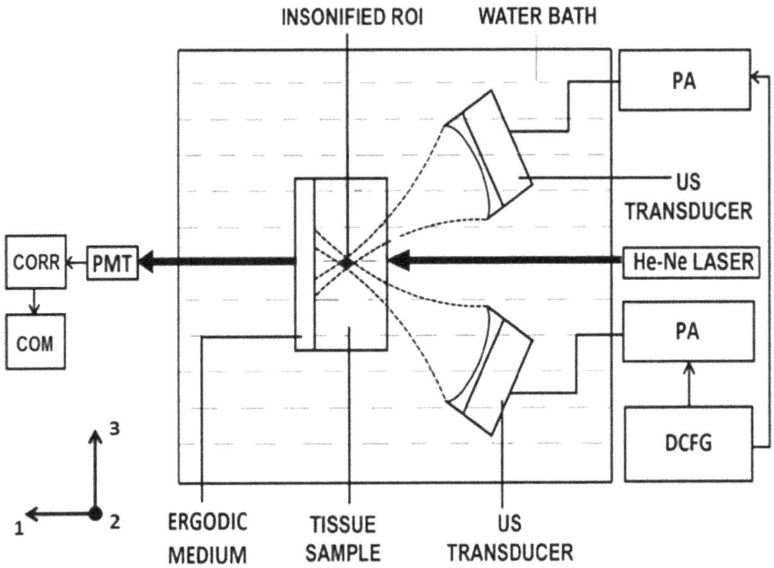

Figure 4.13. Schematics of the experimental setup for ultrasound-assisted DWS. The ergodic medium-object slab composite object, immersed in a water bath (maintained at 27 ± 0.5 °C) is insonified by the intersecting focal volume (ROI) of two ultrasound transducers. The ROI is illuminated by a coherent beam from a He–Ne laser. The scattered light is detected by a single-mode fibre and given to a PMT. The output from the PMT after signal conditioning is input to the correlator and then to the personal computer. PA: power amplifier, US: ultrasound, ROI: region of interest, DCFG: dual channel function generator, CORR: correlator, COM: personal computer. Reprinted with permission from [32]. (2017) Optical Society of America.

fat slab. The transducers are operated, one at the central frequency of 1 MHz and the other at 1 MHz + $\Delta\varpi$ Hz, where $\Delta\varpi$ can be varied from a few hundred Hz to a few kHz. As mentioned in chapter 3, the reason for the use of two transducers to provide 'mixing' at the focal region is to create a low-frequency acoustic wave and through it a radiation force at the difference frequency $\Delta\varpi$.

The object under study is sandwiched to a cuvette containing an ergodic medium, which is a dilute glycerol solution containing, 0.02% by volume 2 μm diameter polystyrene balls. This is done to give ergodicity to the measured signal from the non-ergodic slab of agar or pork fat [33]. As usually done in DWS, we measure $g_2(\tau)$, the intensity autocorrelation. For details of how to extract $g_2(\tau)$ for the slab from that measured for the sandwich and the glycerol solution, see [33].

The aim of the experiment here is to get growth of MSD with delay time of a specified region in the object marked out as ROI by the intersecting ultrasound beams. The autocorrelator, which receives signals from the photon-counting PMT outputs $g_2(\tau)$. (Details of the experiments are already given in chapter 2 (section 2.4.1.1) which also gives the setting used in the PMT (for example, photon-mode) for data gathering so that further processing needed to extract the depth of modulation, $M(\tau)$, overlying $g_1(\tau)$ is easily carried out.) In the second step, $g_1(\tau)$ is computed from $g_2(\tau)$ using Siegert relation. $M(\tau)$ is obtained from $g_1(\tau)$ through its evolutionary power spectrum, as discussed in section 2.3.1. Mean-squared displacement as a function of τ is obtained from $M(\tau)$ using equation (2.16).

We have gathered $g_2(\tau)$ corresponding to three agar samples of increasing stiffness and one pork-fat slab, from which the MSDs are computed. The results from pork-fat slab are plotted in figure 4.14. Results from agar slabs are similar. Figure 4.14(a) (which is the same as that shown in figure 4.8, and repeated here for convenience) shows the effect of the introduction of ultrasound on the fluctuation in the plateau region. Without the ultrasound, forced vibrations in the ROI are not excited (and therefore not amplified by SR) and mixed up with the steady-state dynamics of the Brownian particles seen in the plateau as noise of small intensity. In figure 4.14(b) we have a comparison of the experimental result and that obtained by solving the GLE (equation (4.13)). To solve the GLE we have employed Monte Carlo simulation. The damping memory kernel is approximated by Prony series. Using one of the experimentally measured MSD plots as data, we have used a nonlinear filtering scheme [34] to extract the Prony series coefficients as well as the intensity of the multiplicative noise. With these parameters we have solved equation (4.13) for $\Delta r^2(\mathbf{r}, \tau)$ which is shown in figure 4.14(b) along with its experimental counterpart. The match seems to be quite good.

4.4.3.2 Computation of resonant modes from experimental data and inversion for elastic constants using ensemble Kalman filter

The fluctuations observed in the plateau region of all the four (three for agar and one for pork) MSD plots are subjected to a Fourier analysis for power spectra computation. The obtained power spectra are shown in figures 4.15(a)–(d). In the experiments, we have not optimised the multiplicative and additive noise intensities

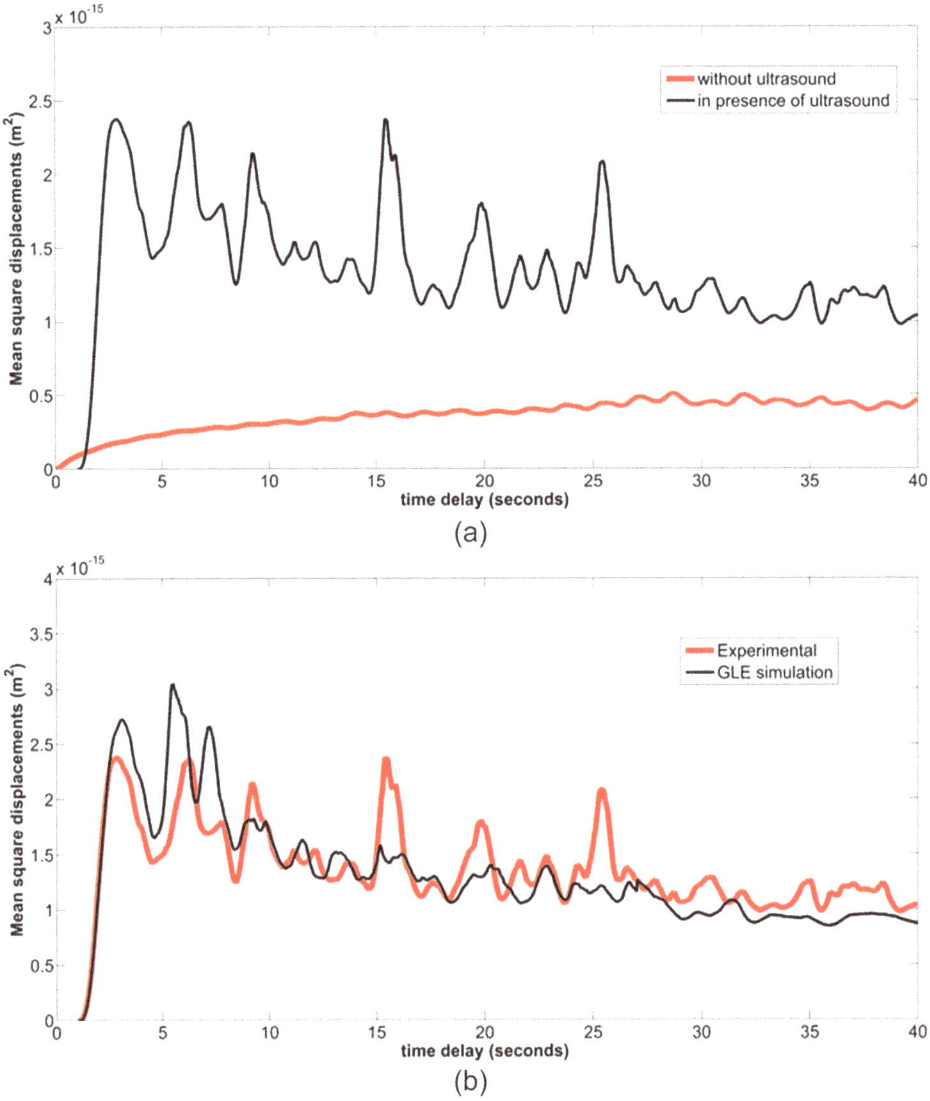

Figure 4.14. (a) The variation of MSD with τ from the measured $g_1(\tau)$ from pork slab. The thin (black) curve corresponds to when the ultrasound is on and the thick (red), when it is off. (b) Comparison of MSD versus τ plot from a pork slab, obtained experimentally (thick (red) curve) with the corresponding one obtained through solving the GLE (equation (4.13)) (thin (black) curve). Reprinted with permission from [32]. (2017) Optical Society of America.

for maximum enhancement of the vibrational modes. Notwithstanding this, with experiments done at room temperature on agar and pork we are able to clearly see four peaks at frequencies, say, ϖ_i, $i = 1, -4$ (see figures 4.15(a)–(d)). Our objective here is to use these resonant frequencies as data and extract the elasticity tensor through an inversion of the 'forward equation' which is the eigenvalue equation, i.e. equation (4.4). If we consider the material of the object as anisotropic the elasticity tensor will have 21 unknown components to be recovered. Since our data is limited

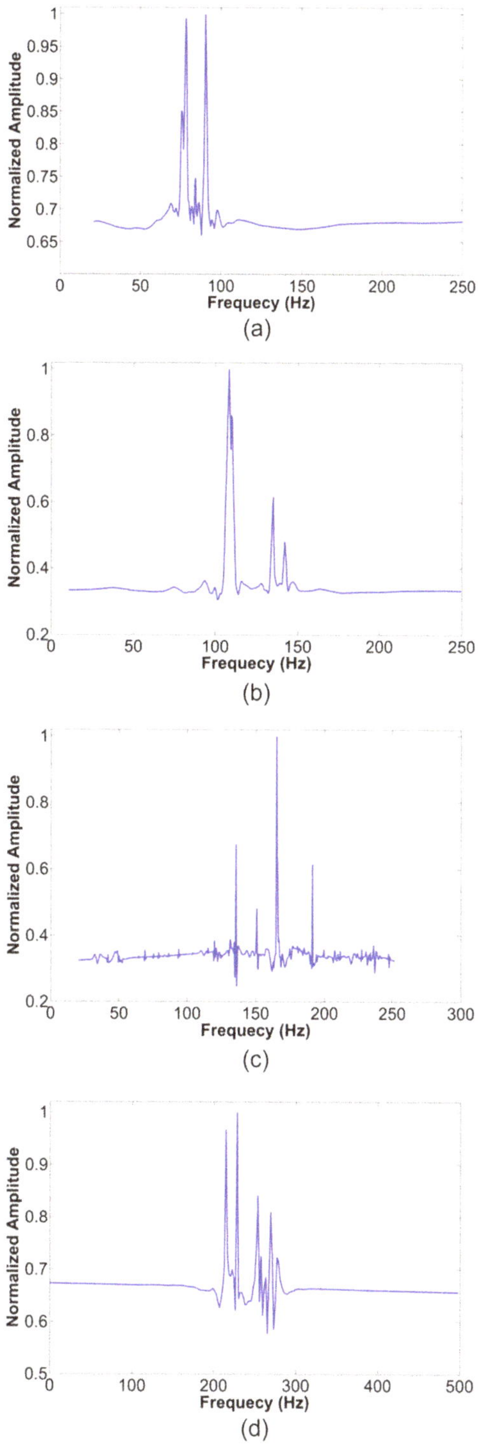

Figure 4.15. (a) Power spectrum obtained from the plateau of the experimentally obtained MSD versus τ curve from 0.5% agar slab, displaying the peaks (resonant modes) at 75.5 Hz, 78.7 Hz, 84.2 Hz and 90.4 Hz; (b) same as (a) but for 1% agar slab. The peaks are at 108.0 Hz, 110.0 Hz, 135.0 Hz and 142.0 Hz; (c) same as (a) but for 1.5% agar slab. The peaks are at 135.8 Hz, 151.0 Hz, 165.0 Hz and 191.4 Hz; and (d) same as (a) but for pork slab. The peaks are at 214.8 Hz, 228.9 Hz, 253.9 Hz and 269.5 Hz. Reprinted with permission from [32]. (2017) Optical Society of America.

to just four, short of tuning the noise intensities which should reveal still higher order modes, we have enforced a symmetry by aligning the edges the object slab along the Cartesian coordinates (1,2,3) (see figure 1 in [32]). With this, we have orthotropy with nine unknowns in the elastic tensor. The forward equation (4.4) is rewritten here for convenience: $\omega^2\mathbf{\Gamma} \cdot \boldsymbol{\alpha} = \mathbf{M} \cdot \boldsymbol{\alpha}$. The elasticity 4-tensor $\mathbf{C} = \{C_{ijkl}\}$ which enters the equation through stiffness matrix $\mathbf{\Gamma}$ will have only nine independent constants under assumption of orthotropy. The stress–strain relation in this case can be written as $\boldsymbol{\sigma} = \mathbf{C}\boldsymbol{\varepsilon}$, where $\boldsymbol{\varepsilon}$ is the small strain tensor. As a matrix–vector equation it can be written in full form as

$$\begin{pmatrix} \sigma_{11} \\ \sigma_{22} \\ \sigma_{33} \\ \sigma_{23} \\ \sigma_{13} \\ \sigma_{12} \end{pmatrix} = \begin{bmatrix} E_1(1-\nu_{23}\nu_{32})\Upsilon & E_1(\nu_{21}+\nu_{31}\nu_{23})\Upsilon & E_1(\nu_{31}+\nu_{21}\nu_{32})\Upsilon & 0 & 0 & 0 \\ & E_2(1-\nu_{13}\nu_{31})\Upsilon & E_2(\nu_{32}+\nu_{12}\nu_{31})\Upsilon & 0 & 0 & 0 \\ & & E_3(1-\nu_{12}\nu_{21})\Upsilon & 0 & 0 & 0 \\ & \textit{symmetric} & & G_{23} & 0 & 0 \\ & & & & G_{13} & 0 \\ & & & & & G_{12} \end{bmatrix} \begin{pmatrix} \varepsilon_{11} \\ \varepsilon_{22} \\ \varepsilon_{33} \\ \varepsilon_{23} \\ \varepsilon_{13} \\ \varepsilon_{12} \end{pmatrix} \tag{4.15}$$

where $\Upsilon = \dfrac{1}{(1-\nu_{12}\nu_{21}-\nu_{23}\nu_{32}-\nu_{31}\nu_{13}-2\nu_{21}\nu_{32}\nu_{13})}$, $\{E_i\}$ are elastic moduli in the directions $i = 1, 2, 3$, $\{\nu_{ij}\}$ are Poisson's ratios along j for strain along i and G_{23}, G_{13}, G_{12} are shear moduli in 2–3, 1–3 and 1–2 planes. Moreover, we have used column vectors for representing stress and strain tensors. The first three components of the stress tensor give longitudinal-stress and the last three shear-stress components. Similarly, for the strain column vector, the first three are axial components and the last three shear.

In the following sub-section we describe the inversion of the eigenvalue equation. One of the computations done repeatedly for this is for resonant modes, given the required material properties and other inputs to equation (4.4), in other words solution of the forward problem. For this we have employed ABAQUS, a standard commercial software. The inputs that are required for ABAQUS modal analysis routine are (i) material properties such as density, elasticity tensor, etc, (ii) geometry of the ROI and (iii) boundary condition, which is Dirichlet, in our case, setting amplitude of vibration to zero on the boundary of the VROI. Before describing the inversion, we report the verification of the experimentally measured natural frequencies of the objects by computing the natural frequencies using the modal analysis routine making use of the properties of the objects from literature. The results are in table 4.3, which are closest to the resonant frequencies experimentally measured. We have noticed that through modal analysis we obtained many more natural modes than are measured through experiments. The reason is that either the ultrasound forcing is unable to excite all the computed modes or that the higher order modes are too weak to be measured experimentally. Double stochastic resonance and AESR provide amplification to all the modes; however, tuning of multiplicative and additive noise strengths to maximize amplification is not done in

Table 4.3. Verification of the experimentally measured resonant modes of the objects through those computed using a modal analysis [15]. For modal analysis, a commercial software ABAQUS was employed. Reprinted with permission from [32]. (2017) Optical Society of America.

Percentage of agar	1st Mode (Hz)		2nd Mode(Hz)		3rd Mode(Hz)		4th Mode(Hz)	
	Abaqus	Experimental	Abaqus	Experimental	Abaqus	Experimental	Abaqus	Experimental
0.5	73.2	75.5	77.4	78.7	84.5	84.2	94.2	90.4
1.0	106.0	108.0	112.0	110.0	122.2	135.0	136.4	142.0
1.5	148.5	135.8	156.6	151.0	170.9	165.0	190.6	191.4
PORK	225.0	214.8	237.5	228.9	260.0	253.9	290.0	269.5

the experiments. It is conjectured that with proper tuning of noise many more modes could be experimentally measured.

A. Inverse problem

To invert the measured eigenvalues or natural frequencies for $\mathbf{M}$and $\mathbf{\Gamma}$which would give us the unknown elastic tensor components we rewrite equation (4.4) as

$$(\omega^2\mathbf{M} - \mathbf{\Gamma})\boldsymbol{\alpha} = \mathbf{0} \tag{4.16}$$

For a nontrivial solution of equation (4.16) we should have

$$\det(\omega^2\mathbf{M} - \mathbf{\Gamma})\boldsymbol{\alpha}) = 0 \tag{4.17}$$

which is a polynomial in ω of degree $2N$, where N is the DOF of the eigenvalue problem stated in equation (4.4). Our objective is to recover the entries that make up $\mathbf{\Gamma}$ from the four measured natural frequencies through the experiments, assuming $\mathbf{M}$ is fully known. We pose this inversion as a stochastic filtering problem and solve it using the ensemble Kalman filter (EnKF) [35]. With the material considered orthotropic there are nine unknowns whose interpretation depends on the selection of coordinate axes. Out of these we concentrate only on six, three Young's moduli and three shear moduli, assuming the Poisson's ratios are known.

Use of the evolutionary stochastic filtering involves defining all the variables, the unknown nine parameters and four measurements, as diffusive stochastic processes in an appropriate probability space. Out of these the unknown elastic constants (say, the vector process, $\mathbf{X}_t$) evolve over pseudo-time. In addition, we have a measurement equation, which takes $\mathbf{X}_t$ as input and output measurements, which are the eigenvalues possibly corrupted by measurement noise (call it $\mathbf{m}_t$). This operation requires the use of eigenvalue solver from ABAQUS. The objective of the evolutionary search scheme is to arrive at the probability density function for $\mathbf{X}_t$ conditioned on $\mathbf{m}_t$, the measurement up to the current time. With this one can estimate the expected value of $\mathbf{X}_t$ (say, $\overline{\mathbf{X}}_t$) which would drive the innovation, (i.e. the difference between experimental and simulated measurements) to a zero-mean

Brownian process. Further details on EnKF, etc, specifically in the context of the present problem, are available in our paper [32].

4.4.4 Results and discussion

We have used the above scheme to recover six unknown components of the elasticity tensor from the measured four modal frequencies using agar and pork-fat slabs. We started with all the nine components as unknowns; however, after a few iterations the three Poisson's ratios became invariant with respect to iterations, having remained at the initial guess for these objects which are nearly incompressible. Thereafter, though we had all the nine components as unknowns, for all practical purposes we had only six, the Young's and shear moduli components. Should we assume full anisotropy for the material, then with 4 measurements and 21 unknowns, in our experience the filtering density tends to acquire too many peaks to have a proper and meaningful recovery of the unknowns. (This brings home the need for increasing the dimension of measurement by maximizing the amplification of weak vibrational modes. In the present work we have seen that there is noise-assisted order creation and enhancement of weak sinusoids; however, further exploration is needed to optimize noise intensities so that the enhancement is maximised.) Even within the orthotropic assumption we have worked with, one needs to enforce certain *a priori* constraints to ensure convergence of the recursion. For example, for physical realizability of the orthotropic parameters, the following constraints have to be met:

$$E_1, E_2, E_3, G_{23}, G_{31}, G_{12} > 0; \frac{\nu_{ij}}{E_i} = \frac{\nu_{ji}}{E_j}\ i, j = 1, 2, 3; \text{ and}$$
$$\nu_{21}\nu_{32}\nu_{13} < \frac{1 - \nu_{21}{}^2\frac{E_1}{E_2} - \nu_{32}{}^2\frac{E_2}{E_3} - \nu_{13}{}^2\frac{E_3}{E_1}}{2} < \frac{1}{2} \tag{4.18a–c}$$

A simple strategy employed in our computations to remain within the constraints is to start the algorithm with the assumption of isotropy leading to quick convergence for three components: Young's and shear moduli and Poisson's ratio. These values are used as an initial guess to start the algorithm for the orthotropic case. Figure 4.16 shows the evolution of Young's moduli along the three directions with recursion number; figure 4.17 is the same for shear moduli. Insets in these figures show the parameters' initial behaviour, with the assumption of isotropy. In particular, we would like to point out the behaviour shown in figures 4.16(d) and 4.17(d), which are obtained from pork fat. It is seen that the converged values are dependent on direction (or on coordinate axes) revealing that pork fat, even though elastically anisotropic strictly, could have orthotropic behaviour. Figure 4.18 shows the photograph of one of the fat slabs used in the experiments showing also in the inset how the coordinate axes were selected. The appearance of a rod-like structure running perpendicular to the surface can be noticed. This particular structure is meant to provide extra strength in the direction marked 1 in the figure. Our algorithm recovered a higher value in this direction for Young's modulus

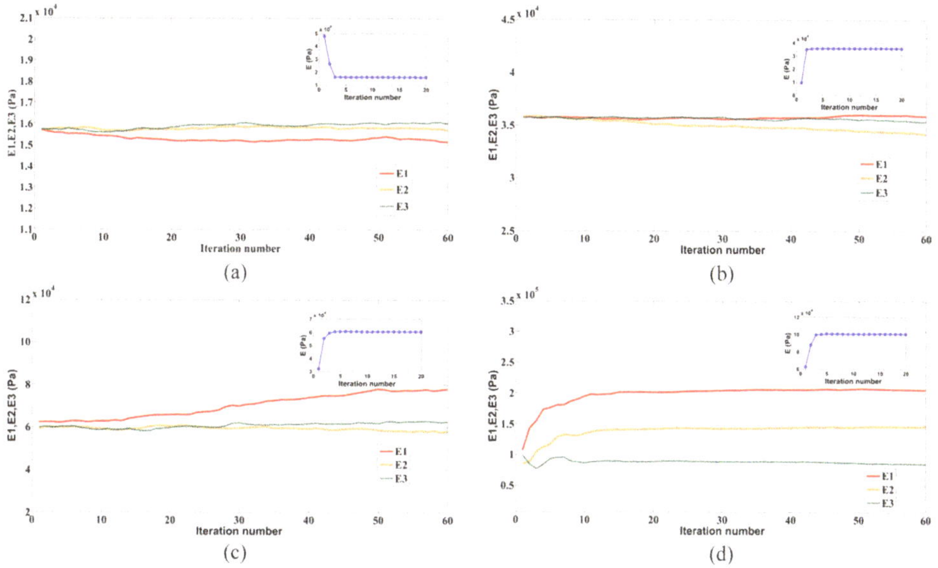

Figure 4.16. The estimated Young's moduli, E_1, E_2, and E_3 from the measured resonant modes for slabs (a) with 0.5% agar, (b) with 1% agar, (c) with 1.5% agar and (d) pork fat. Notice that for pork, being orthotropic, E_1, E_2, and E_3 are different at convergence. For all the four cases, the object was modelled to be linear, orthotropic. Inset shows the initial convergence with assumption of isotropy. Reprinted with permisssion from [32]. (2017) Optical Society of America.

(denoted E_1 which is 205.8 kPa) and the shear modulus associated with this, which is G_{12}=48.1 kPa, compared to those in other directions (figures 4.16 and 4.17). We also note, in passing, that pork fat harvested from near internal organs, and away from the skin is reported to show isotropic behaviour.

Table 4.4 gives a compilation of all the parameters recovered from the experimentally measured resonant frequencies for all the four slabs, three agar and one pork fat. For agar we have comparison with independent measurements using a rheometer; for pork the comparison is with the published result in [36] where the material is assumed isotropic. (We point out that the single Young's modulus form [36] is 140 kPa which is close to one of the values we have reconstructed, 146 kPa.)

Figure 4.19 gives the behaviour of root mean-square error (RMSE) with recursion, the RMSE computed in the measurement domain, when the object is pork fat. The RMSE for recursion k is calculated as $\mathrm{RMSE}_k = \sqrt{\sum_{i=1}^{4}\left(f_i^{\text{experiment}} - \hat{f}_i^{k}\right)^2}$. It is noted that for up to 20 recursions the error was stagnant, after an initial fall which lasted until the third recursion, because we have assumed that the material is isotropic. Beyond 20, we assumed orthotropic assumption, and the error started to fall until it reached sufficiently low values beyond 100 recursions. For comparison, we have also shown here the behaviour of error for a typical agar slab, for which in three recursions

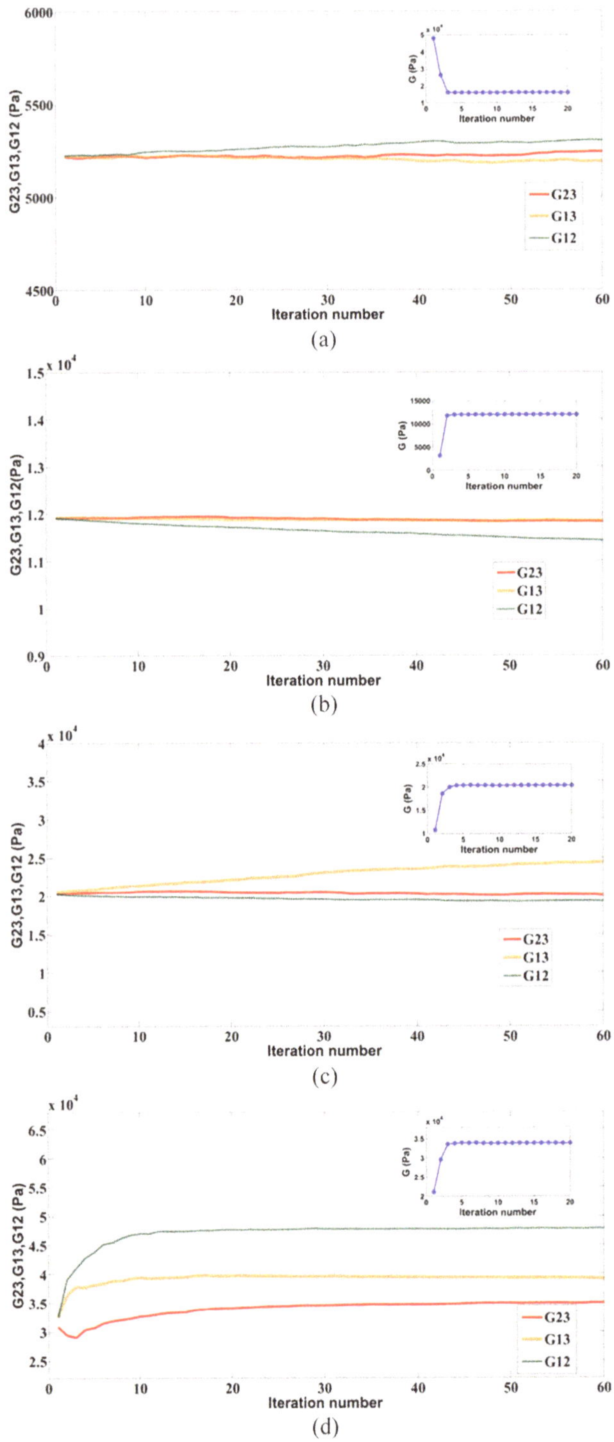

Figure 4.17. Same as figure 4.16 but showing shear moduli (a) with 0.5% agar, (b) with 1% agar, (c) with 1.5% agar, and (d) for pork fat. Notice that for pork, being orthotropic, G_{23}, G_{13} and G_{12} are different at convergence. For all the four cases, the object was modelled to be linear, orthotropic. Inset shows the initial convergence with assumption of isotropy. Reprinted with permission from [32]. (2017) Optical Society of America.

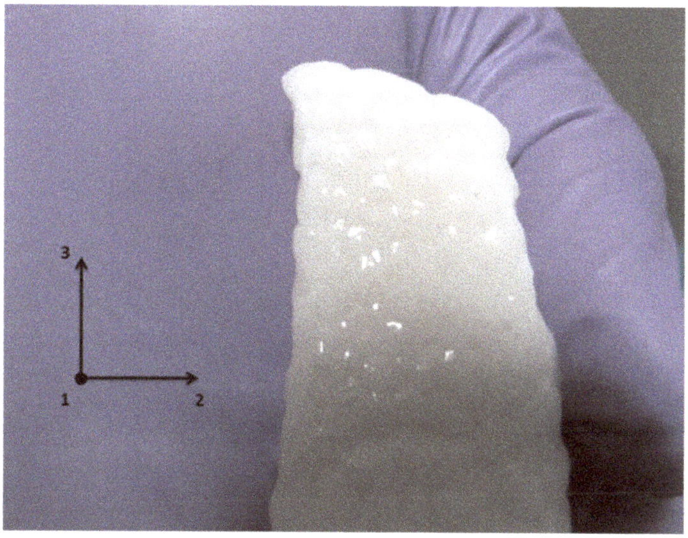

Figure 4.18. Pork back fat used during the experiment with the coordinate axis (inset), representing the material directions. The surface visible in this figure was just below the skin, as can be confirmed from its grainy appearance. Reprinted with permission from [32]. (2017) Optical Society of America.

the error reached acceptably low values and remained so when recursions continued. This is so, because the initial assumption of isotropy is not misplaced.

If the measurement of natural frequencies becomes error prone, quite naturally the recovered elasticity parameters become error prone. We did an analysis of this error, which is reported in table 4.5. It is observed that the error introduced in elasticity parameter estimation is more pronounced owing to error in lower order natural frequencies than higher. Due to a measurement error of 5 Hz in the first natural frequency the error introduced in G_{12} is 10 kPa, whereas the same error in the fourth natural frequency introduced an error of 3 kPa in E_1.

4.5 Concluding remarks

This chapter has two main sections, both of them have the same aim: quantitative vibro-acoustography from the measurement of natural frequencies of vibration from a region excited by radiation force from ultrasound beams (the ROI). One major contribution of the first part is the derivation of source term for the VA wave which is seen to be modulated by the locally generated shear waves carrying information on shear modulus of the ROI. This has helped us explain how shear modulus information is carried to the boundary of the object with the VA wave acting as a carrier. In the experiment we have used a fibre Bragg grating (FBG) sensor to detect the VA wave and delineate the frequencies of the eigenmodes of vibration. Since the higher order modes are weak, the FBG (limited also by a high-frequency cut-off) could detect only the fundamental mode. We have carried out an eigenvalue analysis of the freely vibrating ROI, and from the symmetric eigenvalue equation recovered Young's modulus of the material of the ROI, assuming it to be linear isotropic, using

Table 4.4. Comparison of elastic parameters estimated by the present method with measurements from rheometer or published results, as the case may be. Reprinted with permission from [32]. (2017) Optical Society of America.

	Agar 0.5%		Agar 1.0%		Agar 1.5%		Pork	
	Estimated parameters	Comparison	Estimated parameters	Comparison	Estimated parameters	Comparison	Estimated parameters	Comparison
E_1 (kPa)	15.1	$E = 15.6$ (calculated using G and considering isotropic)	35.8	$E = 35.0$ (calculated using G and considering isotropic)	78.0	$E = 64.1$ (calculated using G and considering isotropic)	205.7	$E = 140.0$ (table 2 of [36])
E_2 (kPa)	15.7		34.1		58.8		146.5	
E_3 (kPa)	16.0		35.3		62.6		85.2	
G_{23} (kPa)	5.2	$G = 5.2$ (figure 4.2(a))	11.8	$G = 11.6$ (figure 4.2(a))	20.0	$G = 21.3$ (figure 4.2(a))	34.9	$G = 46.7$ (table 2 of [36])
G_{13} (kPa)	5.1		11.8		24.3		39.0	
G_{12}(kPa)	5.3		11.4		19.2		47.7	
ν_{23}	0.4748	0.49 (incompressibility assumption)	0.4748	0.49 (incompressibility assumption)	0.4802	0.49 (incompressibility assumption)	0.4777	$\nu = 0.4999$ (table 2 of [36])
ν_{13}	0.4772		0.4772		0.4806		0.4752	
ν_{12}	0.4771		0.4771		0.4781		0.4609	

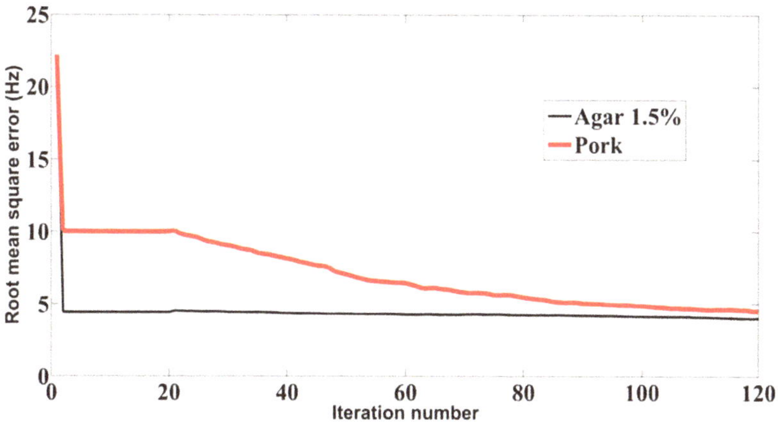

Figure 4.19. Behaviour of root-mean-square error with iteration number during the recovery of the orthotropic elastic parameters of pork fat. During the first 20 iterations the sample was modelled isotropic, and in next 100 orthotropic. Reprinted with permission from [32]. (2017) Optical Society of America.

Table 4.5. Table showing the error in recovered Young's and shear moduli of pork fat owing to error in the measured modal frequencies. Reprinted with permission from [32]. (2017) Optical Society of America.

Mode number	Measured frequency $\Delta\varpi_i$ Hz	$E_1 \pm \Delta E_1$ kPa	$E_2 \pm \Delta E_2$ kPa	$E_3 \pm \Delta E_3$ kPa	$G_{12} \pm \Delta G_{12}$ kPa	$G_{13} \pm \Delta G_{13}$ kPa	$G_{23} \pm \Delta G_{23}$ kPa
1	215 ± 5	205.7 ± 8	146.5 ± 9	85.2 ± 7	47.7 ± 10	39.0 ± 9	34.9 ± 9
2	230 ± 5	205.7 ± 6	146.5 ± 8	85.2 ± 5	47.7 ± 8	39.0 ± 7	34.9 ± 8
3	254 ± 5	205.7 ± 4	146.5 ± 8	85.2 ± 5	47.7 ± 5	39.0 ± 7	34.9 ± 7
4	270 ± 5	205.7 ± 3	146.5 ± 7	85.2 ± 5	47.7 ± 5	39.0 ± 6	34.9 ± 7

the single frequency measurement as data. Since frequency is seldom corrupted by noise (in comparison to amplitude) the recovered results are quite unaffected by noise the VA wave amplitude is quite prone to pick up from the system. Having established a reliable, frequency-based method for doing quantitative vibro-acoustography, the aim of the second part of this study is extending it to measure the elasticity tensor when the material is anisotropic. For this, one needs to enrich the measurement with as many higher order modes of vibration of the ROI as possible. This enrichment is carried out in the second part using DWS as the measurement tool.

In this part, we have explained the amplification of the resonant modes, observable in the fluctuations in the plateau of the MSD plot, using array-enhanced stochastic resonance coupled with doubly stochastic resonance. In the work presented no attempt was made, in the experiments, to tune either the additive noise strength or the multiplicative noise so that the SNR enhancement is maximized. We have carried out the experiments at room temperature and with certain objects, which means at certain additive and multiplicative noise strengths, and found we could measure four modal frequencies. Additive noise can be controlled by varying the strength of the high-frequency interference term as suggested in [28]. In the simulations, it was observed that whereas the strength of additive noise played little role, the multiplicative noise played a significant one in enhancing SNR of the output. With only four frequency measurements we had to assume the material to be orthotropic and limit to the recovery of essentially six unknowns. More investigations and experiments are needed to arrive at a means to tune noise strengths so that many more resonance modes would become visible. If the data is richer, then we can recover elasticity tensor assuming full-fledged anisotropy. Recovery of many more elasticity tensor components from the set of 21 unknowns would be helpful in early diagnosis of diseases such as malignancy; for, such investigations would throw light on which of the constants are more sensitively dependent on progress of malignancy.

References

[1] Zadler B J *et al* 2004 Resonant ultrasound spectroscopy: theory and application *Geophys. J. Int.* **156** 154–69

[2] Barnes J A and Allen D W 1966 A statistical model for flicker noise *Proc. IEEE* **54** 176–8

[3] Sarkar S *et al* 2015 Internal noise-driven generalized Langevin equation from a nonlocal continuum model *Phys. Rev.* E **92** 022150

[4] Gammaitoni L *et al* 1998 Stochastic resonance *Rev. Mod. Phys.* **70** 223–87

[5] McNamara B and Wiesenfeld K 1989 Theory of stochastic resonance *Phys. Rev.* A **39** 4854–69

[6] Lindner J F *et al* 1995 Array-enhanced stochastic resonance and spatiotemporal synchronization *Phys. Rev. Lett.* **75** 3–6

[7] Mazumder D *et al* 2017 Quantitative vibro-acoustography of tissue-like objects by measurement of resonant modes *Phys. Med. Biol.* **62** 107

[8] Fatemi M and Greenleaf J F 1998 Ultrasound simulated vibro-acoustic spectrography *Science* **280** 80–5

[9] Fatemi M and Greenleaf J F 1999 Vibro-acoustography: an imaging modality based on ultrasound-stimulated acoustic emission *Proc. Natl. Acad. Sci.* **96** 6603–8

[10] Mitri F G and Kinnik R R 2012 Vibroacoustography imaging of kidney stones in vitro *IEEE Trans. Biomed. Eng.* **59** 248–54

[11] Brigham J C *et al* 2007 Inverse estimation of visco-elastic material properties for solids immersed in fluids using vibro-acoustic techniques *J. Appl. Phys.* **101** 023509

[12] Aguilo M A *et al* 2010 An inverse problem approach to elasticity property imaging through vibroacoustics *IEEE Trans. Med. Imag.* **29** 1012–29

[13] Zadler B J *et al* 2004 Resonant ultrasound spectroscopy: theory and application *Geophys. J. Int.* **156** 154–69

[14] Migliori A *et al* 1993 Resonant ultrasound spectroscopic techniques for measurement of elastic moduli of solids *Physica* B **183** 1–24

[15] Visscher W M *et al* 1991 On the normal modes of free vibration of inhomogeneous and anisotropic elastic objects *J. Acoust. Soc. Am.* **90** 2154–62

[16] Barnes H A *et al* 1989 *An Introduction to Rheology* vol 2 (Amsterdam: Elsevier)

[17] Hill K O and Meltz G 1997 Fiber Bragg grating technology: fundamentals and overview *J. Lightwave Technol.* **15** 1263–76
[18] Chandran R S *et al* 2014 Diffusing-wave spectroscopy in an inhomogeneous object: local visco-elastic spectra from ultrasound-assisted measurement of correlation decay arising from the ultrasound focal volume *Phys. Rev.* E **90** 012303
[19] Kubo R 1966 The fluctuation-dissipation theorem *Rep. Prog. Phys.* **29** 255–84
[20] Glauber R J 1963 Time-dependent statistics of the Ising model *J. Math. Phys.* **4** 294–307
[21] Bulsara A and Gammaitoni L 1990 Tuning to noise *Phys. Today* **49** 39–45
[22] Chapeau F and Godivier X 1997 Theory of stochastic signal transmission in static nonlinear systems *Phys. Rev.* E **55** 1478–95
[23] Dykman M I *et al* 1995 Stochastic response in perspective *Nuovo Cimento* D **17** 661–83
[24] Jung P 1995 Stochastic resonance and optimal design of threshold detectors *Phys. Lett.* A **207** 93–104
[25] Hu G *et al* 1992 A study of stochastic resonance without adiabatic approximation *Phy. Lett.* A **172** 21–8
[26] Gora P F 2003 Array-enhanced stochastic resonance and spatially correlated noise (arXiv: cond-mat/0308620v1)
[27] Kharchenko V and Gyochuk I 2012 Flashing subdiffusive ratchets in viscoelastic media *New J. Phys.* **14** 043042
[28] Landa P S and McClintock P V E 2000 Vibrational resonance *J. Phys. A: Math. Gen.* **33** 1433–8
[29] Van den Broeck C *et al* 1994 Noise-induced nonequilibrium phase transition *Phys. Rev. Lett.* **25** 3395–8
[30] Zaikin A A 2002 Doubly stochastic effects *Fluct. Noise Lett.* **2** L157–68
[31] Zaikin A A *et al* 2000 Doubly stochastic resonance *Phys. Rev. Lett.* **85** 227–31
[32] Mazumder D *et al* 2017 Orthotropic elastic moduli of biological tissues from ultrasound-assisted diffusing-wave spectroscopy *J. Opt. Soc. Am* A **34** 1945–56
[33] Scheffold F *et al* 2001 Diffusing-wave spectroscopy of non-ergodic media *Phys. Rev.* E **63** 061404
[34] Sarkar S *et al* 2014 A Kushner–Stratonovich Monte Carlo filter applied to nonlinear dynamical system identification *Physica* D **270** 46–59
[35] Roy D and Rao G V 2017 *Stochastic Dynamics, Filtering and Optimization* (Cambridge: Cambridge University Press)
[36] Glozman T and Azhari H 2010 A method for characterization of tissue elastic properties combining ultrasonic computed tomography with elastography *J. Ultrasound Med.* **29** 387–98

IOP Publishing

Ultrasound-Mediated Imaging of Soft Materials

Ram Mohan Vasu and Debasish Roy

Chapter 5

Light diffusion from non-spherical particles: rotational diffusion micro-rheology using ultrasound-assisted diffusing-wave spectroscopy

5.1 Introduction

In the previous chapters, chapters 2 and 4, we have discussed an ultrasound-assisted method to measure local visco-elastic properties, in particular through ultrasound-assisted local diffusing wave spectroscopy (DWS). The principal aim of chapter 4 is exploiting the amplification of the higher order modes of the VROI which DWS provides; recovery of elastic properties is by inverting the eigenvalue equation for the VROI. As discussed in chapter 2, DWS measurements provide an alternative route for recovery of the visco-elastic spectrum of the medium hosting the scattering centres. This recovery used the FDT (fluctuation dissipation theorem) and through it the generalized Stokes–Einstein equation (GSER) to connect a quantity associated with dynamics of the scattering centres, for example the MSD, to the mechanical property of the host medium which in our case is the complex modulus of elasticity. In most of the micro-rheological applications of DWS, scattering particles are assumed to be spherical which grants them scattering isotropy; which means that the phase fluctuation we measure through DWS does not have contributions from rotational Brownian motion of the scattering centres [1]. In most biological objects, the scattering centres have shape anisotropy with consequent scattering anisotropy. Even in such objects with marked anisotropy in scattering, if the object volume, V, is much larger compared to $(l^*)^3$, the object has isotropic scattering with length-scale of l^* and scattering coefficient of μ_s'. This would mean that scattered light reaching the detector will have phase owing to micro-rotations averaged out, and both directions of scattering and information on rotational diffusion are simply absent in the phase fluctuations computed through diffusive light-scattering experiments. As noted in [2], there is an early time-window in which anisotropic *particle* diffusion is observed and a crossover to isotropic diffusion as time progresses. If a rotational diffusion coefficient, D_θ, can be assigned for 3D rotational diffusion, then time $\tau_\theta = \frac{1}{3D_\theta}$ sets the limit of the time window within which anisotropic diffusion is observed. Light scattered from anisotropic scattering centres undergoing temperature-induced Brownian motion shows so-called transport behaviour within a few initial scattering events (number of events depends on the anisotropy factor, g) and thereafter, having lost information on initial direction of light, diffusion. This initial distance over which there is anisotropic light scattering is the transport mean-free-path, l^*, which could be related to τ_θ the edge of the time-window beyond which particle diffusion starts. As mentioned earlier, in the DWS experiment, owing to the size of the object, we have apparent isotropy in scattering and one does not have information on phase introduced by micro-rotation. But the introduction of acoustic radiation force through focused ultrasound beams and the marking out of an ROI as small as hundreds of μm^3 by mixing two focal regions appropriately, help one isolate photons carrying phase fluctuation from the Brownian motion pertaining to the ROI alone. The ROI being small, the scattering has anisotropy, and the component of phase fluctuation owing to particle rotation is not yet averaged out. Therefore, ultrasound-assisted DWS provides a means to measure rotational diffusion of scattering particles through the measurement of this component of phase fluctuation. As noted in [2], a quantitative assessment of rotational Brownian motion is not easy; in both [1, 2] methods based on direct particle tracking over periods of time have been implemented, in [1] through light-streak tracking and in [2] using video-frames of particles imaged by a CCD camera. In both, ergodicity is assumed to equate time averages to ensemble averages. While this is true for the liquid host medium used in [1, 2], it is not so when the liquid is replaced by a visco-elastic jelly, where ergodicity has to be imposed through the now famous double-cell technique. In the experiments reported here we used tissue or tissue-like material and ergodicity was imposed.

In addition, data from the experiments here contain a large cache of other useful information. For example, transition from anisotropic to isotropic particle diffusion can be tracked by following how anisotropy in light scattering transitions to isotropic scattering, as the volume of the ROI is gradually increased. (This is not part of the experimental investigation reported here, but is left for a future investigation.) Parameters such as shape and size of the particles, multiplicative noise strength which represents nonlocal interaction between particles and complex modulus of elasticity from mean-squared rotation can also be recovered from the overall MSD, i.e. with contributions from linear displacement and rotation. As we shall see in section 5.4, the MSD versus time-delay plot has two parts: the short-time transient giving the rise of MSD with time and the dynamic equilibrium state at larger times which we denote the plateau. As mentioned in chapter 2, since the modulation-depth decay is owing to Brownian motion of particles within the ROI, the rise in the transient part is through response of the particles to pure Brownian thermal forces, largely unaffected by the deterministic forcing from ultrasound. This is not so in the plateau region, where there is a mixing of the random dynamics with the modes of vibration of the ROI which are activated by the acoustic forces. We believe mixing arises because the response of the material of the ROI is pushed from linear to nonlinear by the application of acoustic force that is orders of magnitude larger than the thermal forcing, which stretches the 'spring-and-dashpot' connections between particles to the fullest. Fluctuations in the plateau are more pronounced when the particles have shape asymmetry and the power spectrum thereof contains more natural frequencies of the VROI than possible from similar experiments using spherical scattering centres (see section 5.5.2).

In what follows we describe and analyse the experiments we did with agar phantoms, with isotropic scattering centres as well as anisotropic, and pork fat. Prior to that, in section 5.2, we derive an expression for phase fluctuations when contribution from rotation is not averaged out. The theoretical modelling is further strengthened by the generalized Langevin equation (GLE) describing the dynamics of a representative system particle (section 5.3). Solving the GLE helps us compute variation of MSD with time; we also describe the means to extract the shape parameter of the particles from the MSD curve. The experiments are described in section 5.4. We have gathered intensity autocorrelation from both agar phantoms and pork. After extracting decay of $M(\tau)$, we have recovered the complex modulus of elasticity in respect of all the objects. Verifying these results through independent rheometer measurements, it is noticed that whereas there is a perfect match of elasticity spectra for agar with spherical particles, it was approximately ~1.7 times for agar with non-spherical particles. This is owing to the fact that we have used phase fluctuation data with contributions from translatory as well as rotatory dynamics in the GSER derived for purely translatory dynamics. We have then employed the results of section 5.2 to separate the contributions from rotation in the MSD plot. Then the extracted visco-elastic spectrum matched the one obtained through independent rheometer measurements. Data from pork fat was also inverted for the visco-elastic spectrum.

5.2 Light scattering from an ensemble of particles with shape anisotropy

Our objective is to measure the decay of the field autocorrelation of the ultrasound-tagged photons, which is the same as the decay of $M(\tau)$, the modulation on $g_1(\tau)$. The measurement *per se* is $g_2(\tau)$. We use an expression for $M_s(\tau)$, the modulation decay for photons of pathlength s and ensemble average $M_s(\tau)$ with the probability density function for photon path to get $M(\tau)$. The expression for $M_s(\tau)$ is $M_s(\tau) = \frac{\langle |E_s|^2\rangle}{\langle I\rangle}\langle \exp[j\Delta\varphi_s(\tau)]\rangle$. Here, $\langle I_s\rangle \equiv \langle |E_s^2|\rangle$ is the average intensity of photons that took a path of length s, $\langle I\rangle$ is the average intensity at the detector from all photon paths and $\Delta\varphi_s(\tau)$ is the change in phase of light in time delay τ through photon-paths of length s. Since $\Delta\varphi_s(\tau)$ is a Gaussian random variable,

$$\langle \exp[j\Delta\varphi_s(\tau)]\rangle = \exp[-\langle \Delta\varphi_s(\tau)^2\rangle/2] \tag{5.1}$$

We now derive an expression for $\langle \Delta\varphi_s(\tau)^2\rangle$ which incorporates fluctuations due to translation as well as rotation. Figure 5.1 gives a representation of the photon path

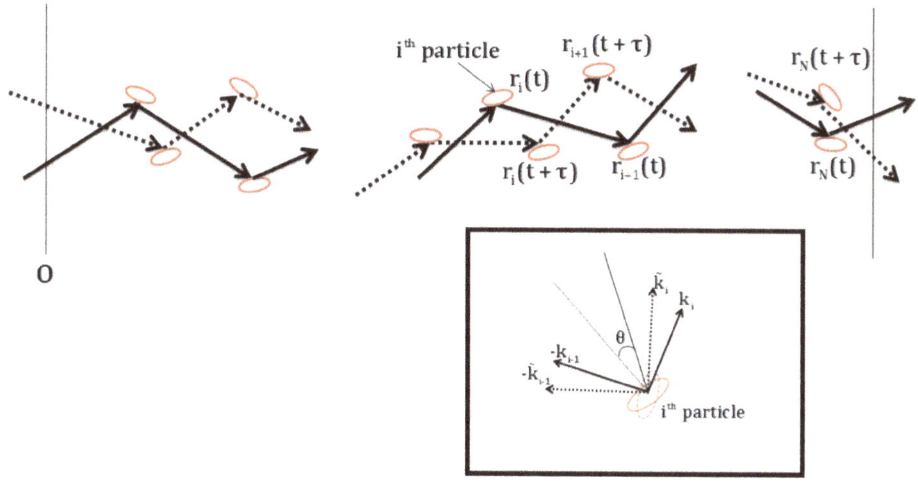

Figure 5.1. A schematic diagram of a typical photon pathway at two successive time instants within the common ultrasound focal volume.

through the object at time t and $t + \tau$. In τ secs the anisotropic scattering centres *en route* to the detector have suffered temperature-induced linear displacements and rotations. (The domain of interest is only the ROI, where the ultrasound focal volumes intersect, providing a low-frequency acoustic radiation force driving the particles to periodic motion, over and above the temperature-induced Brownian motion. Since the correlation decay in $M(\tau)$ is connected to phase fluctuation produced by the Brownian motion, we do not consider the periodic movement of the particles in the ROI.) The particles which typically represent the tissue scattering centres have a forward directed scattering anisotropy with the factor, $g \equiv \langle \cos\theta \rangle$ greater than 0.9. Therefore, the ith scattering centre shown in the inset of figure 5.1 receives the incoming light along a new direction, $\tilde{k}_{i-1}$ (the scattering wave vector from the $(i-1)$th scattering centre after rotation) and scatters light along $\tilde{k}_i$, the directions being mean taken with respect to the scattering phase function. The phase of light reaching the detector at time t sec following a typical path of length s is given by

$$\varphi_s(t) = \mathbf{k}_0 \cdot (\mathbf{r}_1 - \mathbf{r}_0) + \mathbf{k}_1 \cdot (\mathbf{r}_2 - \mathbf{r}_1) + \ldots \mathbf{k}_i \cdot (\mathbf{r}_i - \mathbf{r}_{i-1}) + \ldots \mathbf{k}_n \cdot (\mathbf{r}_d - \mathbf{r}_n) + \varphi_0 \quad (5.2)$$

where φ_0 is the initial phase of light entering the ROI, $\mathbf{r}_i$ is the position vector of the ith scattering centre and $\mathbf{k}_i$ is the scattering-wave propagation vector from it. At time $(t + \tau)$ sec, when the scattering centres have undergone translational and rotational movements, the phase along the new path is

$$\begin{aligned}\varphi_s(t+\tau) = {} & \tilde{\mathbf{k}}_0 \cdot (\mathbf{r}_1 + \Delta\mathbf{L}_1 - \mathbf{r}_0) + \tilde{\mathbf{k}}_1 \cdot (\mathbf{r}_2 + \Delta\mathbf{L}_2 - (\mathbf{r}_1 + \Delta\mathbf{L}_1)) \\ & + \ldots \tilde{\mathbf{k}}_i \cdot (\mathbf{r}_{i+1} + \Delta\mathbf{L}_{i+1} - (\mathbf{r}_i + \Delta\mathbf{L}_i)) \\ & + \ldots \tilde{\mathbf{k}}_n \cdot (\mathbf{r}_d - (\mathbf{r}_n + \Delta\mathbf{L}_n)) + \varphi_0\end{aligned} \quad (5.3)$$

Here, $\tilde{\mathbf{k}}_i$, etc, are the new scattering wave-vectors with $\tilde{\mathbf{k}}_i = \mathbf{k}_i + \Delta\mathbf{k}_i$, $i = 1, 2, \ldots n$. The phase-change in τ sec is

$$\Delta\varphi_s(\tau) = -\sum_{i=1}^{n} \Delta\mathbf{L}_i \cdot [(\mathbf{k}_i - \mathbf{k}_{i-1}) - (\Delta\mathbf{k}_i - \Delta\mathbf{k}_{i-1})] \tag{5.4}$$

which can be written as

$$\Delta\varphi_s(\tau) = -\sum_{i=1}^{n} [\Delta L_i \cdot (K_i - \Delta K_i)] \tag{5.5}$$

Here, $\Delta\mathbf{L}_i$ is the displacement suffered by the ith scattering centre in τ secs.

If there is no rotation, or if the particles have shape isotropy that the angle of scattering is invariant to rotation, then equation (5.5) becomes

$$\Delta\varphi_s(\tau) = -\sum_{i=1}^{n} [\Delta\mathbf{L}_i \cdot \mathbf{K}_\mathrm{i}] \tag{5.6}$$

Therefore, if we do not consider rotation, using equation (5.6) one can show that [3, 4]

$$\langle \exp[j\Delta\varphi_s(\tau)] \rangle = \exp[-\langle \Delta\varphi_s(\tau)^2 \rangle /2] = \exp\left[-\frac{1}{3}k_0^2 < \Delta L^2(\tau)\frac{s}{l^*}\right] \tag{5.7}$$

From equation (5.7) $M(\tau)$ is computed as

$$M(\tau) = \int_0^\infty p(s)\exp\left[-\frac{1}{3}k_0^2 < \Delta L^2(\tau)\frac{s}{l^*}\right]ds. \tag{5.7a}$$

(Here, $\langle \cos\theta \rangle \equiv g$, the scattering anisotropy factor, depends on the shape anisotropy of the scattering particles.)

When rotation is sensed the extra term in equation (5.5) is to be accounted for; which means equation (5.7) will have another term, which, as we shall see, represents the mixing of linear displacement with rotation. This is the extra phase picked-up by the rotated scattered beams from ΔL. This phase-change is $\delta\varphi_s(\tau) = -\sum_{i=1}^{n}[\Delta\mathbf{L}_i \cdot \Delta\mathbf{K}_\mathrm{i}]$. Writing $\Delta\mathbf{K}_i$ as $\Delta\mathbf{K}_i = |\Delta\mathbf{K}_i|\Delta\hat{\mathbf{K}}_i = \theta_r|\mathbf{K}_i|\Delta\hat{\mathbf{K}}_i$ (where, $\Delta\hat{\mathbf{K}}_i$ is the unit vector in the direction of $\Delta\mathbf{K}_i$ and θ_r is the rotation suffered by the scattered beam) one can easily show [3, 4] that

$$\langle \exp[j\delta\varphi_s(\tau)] \rangle = \exp[-\langle \delta\varphi_s(\tau)^2 \rangle /2] = \exp\left[-\frac{1}{3}k_0^2\langle \Delta L^2(\tau)\rangle\langle \theta_r^2(\tau)\rangle\frac{s}{l^*}\right] \tag{5.8}$$

In equation (5.8) l^* becomes l_s, the scattering mean-free-path when photon transport is considered within the ROI and s is the photon pathlength intercepted by the ROI. In the present work we chose to neglect the fluctuations from mixing of MSD and angle of rotation. One of the reasons for this is that the overall effect of this term, being the product of the square of two small quantities, can be negligibly small. However, there is a contribution from micro-rotations to $\langle \Delta\varphi_s^2(\tau) \rangle$ which we need to incorporate whilst we are still within the anisotropic scattering regime within the ROI. To explain this we take the hypothetical case when the translatory motion is absent. With only rotation, phase change owing to rotation by θ_i of the ith scattering centre is $k_0\theta_i x_i$, where x_i is the distance travelled by light between the ith and $(i+1)$th scattering events. The overall phase change $\Delta\varphi_s^r$ along path of length s due to rotation is given by $\Delta\varphi_s^r = \sum_{i=1}^{n} k_0\theta_i x_i$. Therefore, an expression for $\langle(\Delta\varphi_s^r)^2\rangle$ is given by

$$\left\langle(\Delta\varphi_s^r)^2\right\rangle = k_0^2\sum_{i=1}^{n}\langle\theta_i^2\rangle l_s^2 = k_0^2\langle\theta_i^2\rangle l_s^2 n \tag{5.9}$$

(Here, we assume that $\{\theta_i\}'s$ are uncorrelated.)

Here, l_s is the scattering mean-free-path. We also note that light follows the angle of rotation within the ROI with an uncertainty assigned by the scattering anisotropy factor. The overall decay in correlation when both translation and rotation are considered is governed by

$$\exp\left[-\frac{1}{3}k_0^2\left\{\langle\Delta L^2(\tau)\rangle\frac{s}{l_s} + \frac{3}{2}\langle\theta_r^2(\tau)\rangle l_s s\right\}\right] \tag{5.10}$$

From equation (5.10) taking the expectation with respect to pathlength probability density, we arrive at $M(\tau)$, i.e.

$$M(\tau) = \int_0^\infty p(s)\exp\left[-\frac{1}{3}k_0^2\left\{\langle\Delta L^2(\tau)\rangle\frac{s}{l_s} + \frac{3}{2}\langle\theta_r^2(\tau)\rangle l_s s\right\}\right]ds \tag{5.11}$$

Our objective in the present work is to show that the decay in modulation as measured by ultrasound-assisted DWS has indeed contributions from linear as well as rotational Brownian motion. We show, through the experiments below, that, with non-spherical particles, the computed MSD, ignoring rotation, is indeed erroneous, as revealed by the computed $|G^*(\omega)|$. In section 5.3 below we present our numerical simulations to arrive at the MSD of a representative system particle in the ROI. The numerical model uses GLE to model its dynamics, as described in chapter 4.

5.3 Numerical simulation of the particle dynamics

The evolution of the single, predominant translational degree of freedom (DOF) of the system particle is captured by the GLE. The way we arrived at the GLE which incorporates a multiplicative noise to take into account micro-polar rotations (i.e. other DOFs) and nonlocal interactions is fully described in section 2.2.2. In the present work we make use of the amplitude of the multiplicative noise term to characterize the particles for shape-anisotropy. In addition, we expand the dissipative kernel in the viscous drag term in the GLE (see equation (2.1)) using Prony series, and use some of the coefficients of expansion to characterize the size of the scattering particles. We also note that, for biological objects whose scattering centres are fat globules and mitochondria, this morphological characterization has diagnostic utility. This noise is derived in [5] and is shown to be zero mean Gaussian and its strength is proportional to the weighted sum of micro-rotations suffered by the bath particles in the ROI.

We now proceed to integrate the GLE and arrive at the MSD suffered by the system particle. The GLE is a stochastic differential equation (SDE), for the integration of which we have employed the scheme developed in [6] which uses the extended variable formulation. With positive Prony series representation of the damping memory kernel we avoided the first convolution. The second one, which has $W_s^t(t)$ (equation (2.1)) as the kernel and which generates the multiplicative noise, is avoided by assuming $W_s^t(t)$ to be delta-correlated. The convolution terms which can render the SDE non-Markovian are in effect evaluated without time-shift. (Note: the assumption of delta-correlated noise kernel need not be true in general; correlation structure of the noise, which depends on the visco-elastic property of the material of the object, has to be arrived at for each class of materials.) Prony series expansion for the memory kernel is

$$\eta(t) = \sum_{k=1}^{N_k} \frac{c_k}{\tau_k} \exp\left[-\frac{t}{\tau_k}\right], \quad t > 0 \tag{5.12}$$

The number of terms, N_k, used in the simulations here is 4. The τ_k's are relaxation time-constants and $c_k's$ are the coefficients of expansion. We call $\dot{x}(t)$ by $V(t)$ and use the extended variable, $Z_{i,k}(t)$ for the convolution of the ith component of $V(t)$by the kth term of the Prony series in equation (5.12). Then the ith component of $V(t)$ can be written as

$$m dV_i(t) + \omega^2 x_i(t)dt + \sum_{i=1}^{N_K} Z_{i,k}(t)dt = \xi_i(t)dt + N_A B_{i,k}^{n} x_i(t)\sqrt{t}\,dt \tag{5.13}$$

The last term on the right hand side of equation (5.13) is the Brownian standard deviation term by assuming that the multiplicative noise with origin in the micro-polar rotation is delta-correlated, as explained in [5]. Also N_A is the amplitude of the multiplicative noise used.

Following the procedure described in [6] we have integrated equation (5.13) and obtained realizations of $x(t)$ and evaluated the MSD from them.The MSDs computed for a typical agar slab first for spherical and then for non-spherical particles in the ROI to match the experiments described in section 5.4 below, are shown in section 5.5. It is observed that the transient part of the MSD versus time curve for non-spherical particles has a higher slope than its spherical particle counterpart, in addition to its plateau being more intense in fluctuations. The transient parts are used to compute $|G^*(\omega)|$. As seen in section 5.2 the mean-squared fluctuation versus time computed using data from slabs with non-spherical particles contain the rise and fluctuations due to linear displacement and rotation. In this case, we used the arrived at MSD from a similar slab with spherical particles (which contains only variance of linear displacements) to get the variance of $\theta(t)$. From this, using GSER for rotational diffusion [1, 2], we have computed $|G^*(\omega)|$ which matched well with that obtained from the linear MSD, for the isotropic agar slab. Results are further discussed in section 5.5.

We now proceed to describe our experiments. Since ultrasound-assisted DWS has already been explained in chapters 2 and 4, our task here is a simple one: refer the reader to experimental sections of these chapters for details.

5.4 Experiments

The experimental set-up is identical to the one described in section 4.4. The objects used are slabs of dimension 65 mm × 17 mm × 18 mm made of either agar or pork fat. One made of agar is designed to have a storage modulus of 15 kPa, and with scattering provided by embedded polystyrene balls. One set of slabs will have spherical balls of average diameter 3 μm (Sigma Aldrich) and the other set ellipsoidal ones of average major and minor diameters of 3.2 μm and 2.8 μm (Magsphere Inc.), respectively. (See figure 5.2 for images from electron microscope.) For pork fat slab, freshly harvested pieces were preserved in PBS solution at a pH of 7.4. Before the experiments the required slabs were trimmed to size. Data from the experiments are intensity variation with respect to time, detected by the PC-PMT, from which temporal autocorrelation is found with the hard-wired autocorrelator. Intensity data collected from the visco-elastic slab does not make up an ergodic process, but can be made one by sandwiching a cuvette containing a liquid with scattering centres with the slab. Temporal intensity autocorrelation, $g_2(\tau)$, from the autocorrelator is taken as the autocorrelation computed across the sample-space. From $g_2(\tau)$, modulus of the amplitude autocorrelation, $|g_1(\tau)|$, is obtained using Siegert's relation. There is an almost sinusoidal modulation running through the autocorrelations; the depths of modulation ($M(\tau)$) across segments of appropriate time window are computed through windowed Fourier transform. From the experiments we have gathered $M(\tau)$ versus τ data from the three objects, namely (1) agar slab with spherical scattering centres, (2) agar slab with non-spherical scattering centres and (3) pork-fat slab. We invert $M(\tau)$ and arrive at $\Delta L^2(\tau)$ using equation (5.7*a*), which is the appropriate equation when, for agar, scattering centres are spherical. If we used data obtained from non-spherical scattering centres, and arrived at $\Delta L^2(\tau)$ using equation (5.7*a*), i.e. without incorporating phase fluctuations from rotational diffusion, as shown included in equation (5.11),we see a larger growth of MSD in its initial rise and also stronger fluctuations in the plateau region. These results are discussed in the following section.

5.5 Results and discussions

5.5.1 Recovery of shape and size parameters

We first describe the inversion of Prony series coefficients and the amplitude of the internal noise term using equations (5.12) and (5.13). These can be related to the shape-(anisotropy) and size parameters of the scattering centres. For doing this inverse problem we used the stochastic search scheme [7]. The parameters to be recovered are together represented by a vector stochastic process which is additively updated over time recursions, with the objective of driving the error, i.e. the measurement-prediction misfit, to a zero-mean Brownian process. The parameters to be recovered are: (1) the Prony series coefficients, c_k and τ_k, (2) the amplitude of the multiplicative noise term, N_A. For a recursive solution of the inverse problem, we apply the prediction-update strategy to a set of realizations of the parameter (vector) random variable, $\mathbf{p} \equiv \{c_i, \tau_i, N_A\}_{1 \leqslant i \leqslant N_K} \in \mathbb{R}^{2N_K+1}$. The measurement is the experimentally obtained MSD variation, and the prediction is obtained by solving the GLE using the

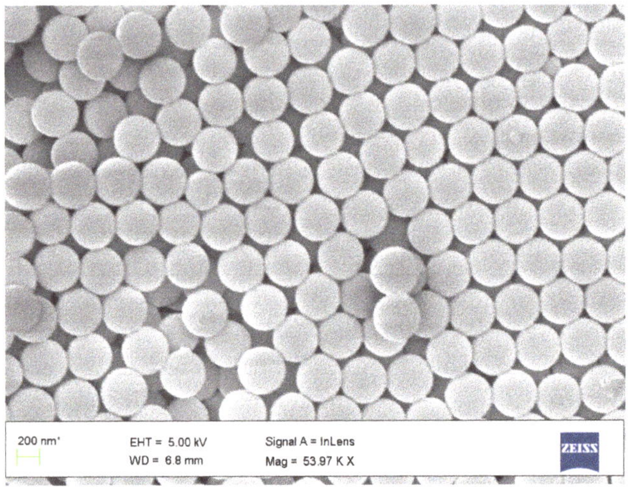

(a)

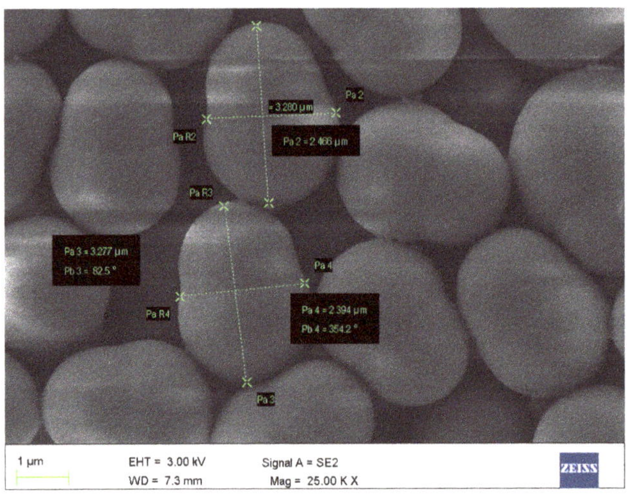

(b)

Figure 5.2. Scanning electron microscope (SEM) images of (a) spherical and (b) non-spherical, ellipsoid polystyrene beads used in the experiments.

integration scheme described in chapter 2, section 2.2.3. The details of the procedure, established for a different application, is available in [7, 8] and therefore not repeated here.

5.5.2 Recovery of mean-squared displacement and visco-elastic spectrum

We first discuss the results of multiplicative noise amplitude and Prony series parameters recovered. As mentioned earlier, the length of the series was limited to four, which means there are eight parameters $\{c_k, \tau_k\}$ and with N_A there are altogether nine unknowns. We have started the stochastic search procedure, with an initial guess of them which was guided by data on average diameter of the spherical particles and the micro-polar theory used to construct the multiplicative noise. The criterion for stopping the algorithm was temporal stability of the drift, which means the norm of the difference in the drift term as recursion proceeds within a preset limit. The recursions behaved well and the parameters recovered for the three objects are shown in table 5.1. Also shown in the table is $\sum_i c_i$ which we conjecture to be proportional to the average diameter of the scattering particles. This, for non-spherical particles was 0.85, and for spherical ones 0.9, and for pork 1.4. From the known value of the average diameter of the particles we found the constant of proportion was 3.53, and 3.32 for the spherical and non-spherical particles, respectively. For animal fat this proportionality constant was found to be 71.4. Using this, the average radius of the scattering centre is seen to be approximately 100 μm which corresponds to the average fat-cell diameter in pork fat from the hind-side. Had we chosen 1 as the proportionality constant, then the average radius would be around 0.25 μm which matched the average radius of mitochondria or cell organelle within the fat cell. We used the fat-cell radius in the GSER to arrive at the complex modulus of elasticity of the material of the slabs.

Comparison of experimental MSD plots, in respect of the three objects, is in figures 5.3–5.5. In addition, figure 5.6 shows comparison of the experimentally obtained MSD plots from agar slabs containing (a) spherical and (b) non-spherical particles. Two aspects stand out: with non-spherical particles, first, the plateau has more fluctuations and second, the rise of the MSD with time is sharper in the initial transient region.

Towards retrieving the visco-elastic spectrum of the material from the early rising part of the MSD curves, we have first dealt with the data presented in figure 5.3. Since the particles have shape symmetry the phase fluctuations are owing only to translatory motion. Therefore, in equation (5.11) there is no contribution from $\langle \theta_r^2(t) \rangle$. Consequently, the right expression to use to recover $|G^*(\omega)|$ is the GSER for linear-diffusion micro-rheology. The expression is given in chapter 2, equation (2.21). Recovered $|G^*(\omega)|$ is plotted in figure 5.7 which closely follows independent measurement using a rheometer. The blue line in the figure, which follows the shape of earlier inversion shown but offset by a factor of ~1.7, is the recovery using linear diffusion GSER of $|G^*(\omega)|$ using data from agar slab with non-spherical particles (figure 5.6, red curve). The discrepancy is because the phase fluctuations measured have contribution from micro-rotations as well, which is not accounted for in the GSER used. This certainly proves that ultrasound-assisted DWS, when the ROI is

Table 5.1. The recovered parameters after solving the inverse problem with EnKF.

	C1, τ1	C2, τ2	C3, τ3	C4, τ4	N_A
Agar with spherical particles	0.17, 2.1	0.25, 3.5	0.07, 8.46	0.36, 8.94	0.5
Agar with non-spherical particles	0.23, 2.16	0.18, 3.3	0.1, 6.54	0.39, 6.7	3.2
Pork	0.2, 3.4	0.31, 3.61	0.54, 4.56	0.35, 9.1	1.3

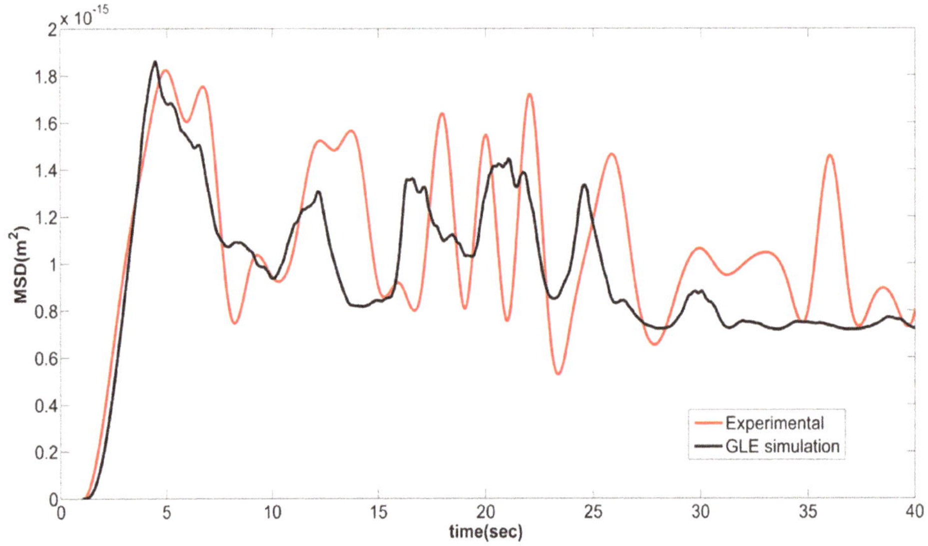

Figure 5.3. Comparison of MSD versus τ plots obtained from experiment (red) with that from solving the modified GLE (black). This is for spherical particles in agar phantom.

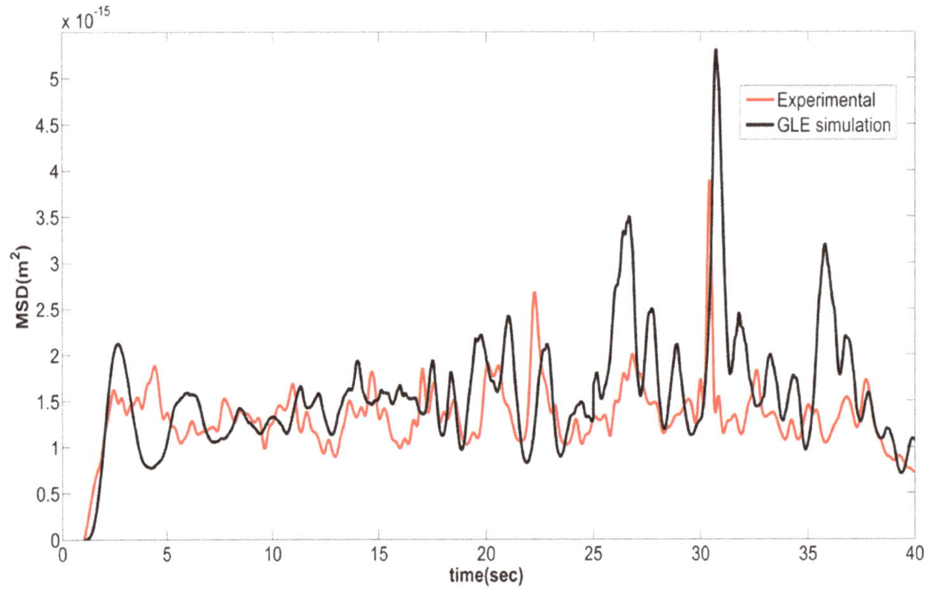

Figure 5.4. Same as figure 5.3 but pertaining to the phantom with non-spherical scatterers.

small, is capable of responding to rotational diffusion of scattering centres in the ROI, through a nonzero measurement of average phase fluctuation. This is due to angular deviation of scattered light from rotation of particles orchestrated by shape anisotropy of the scattering centres. The scattering should not crossover to isotropy

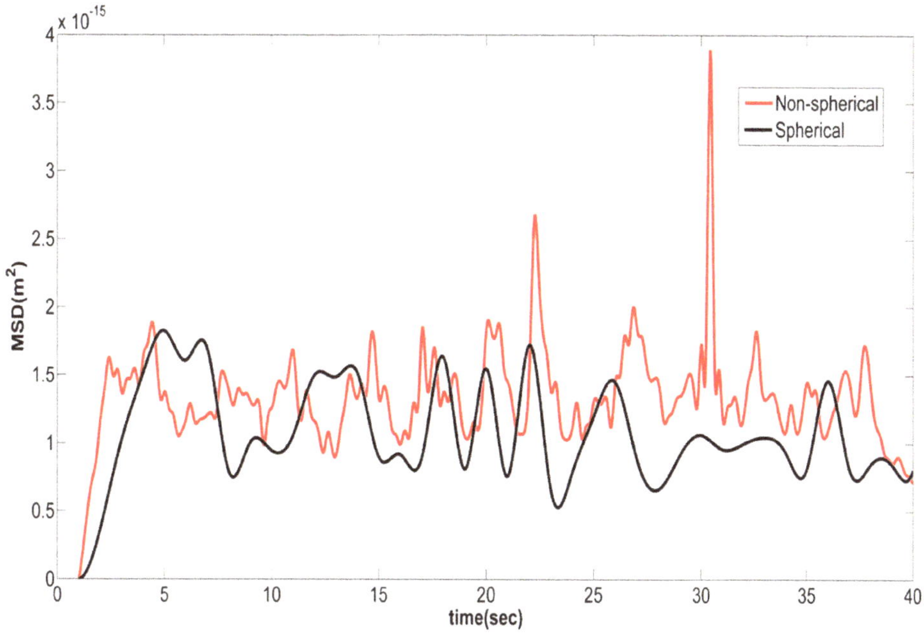

Figure 5.6. Comparison of experimentally obtained MSD versus time plots from agar slabs.

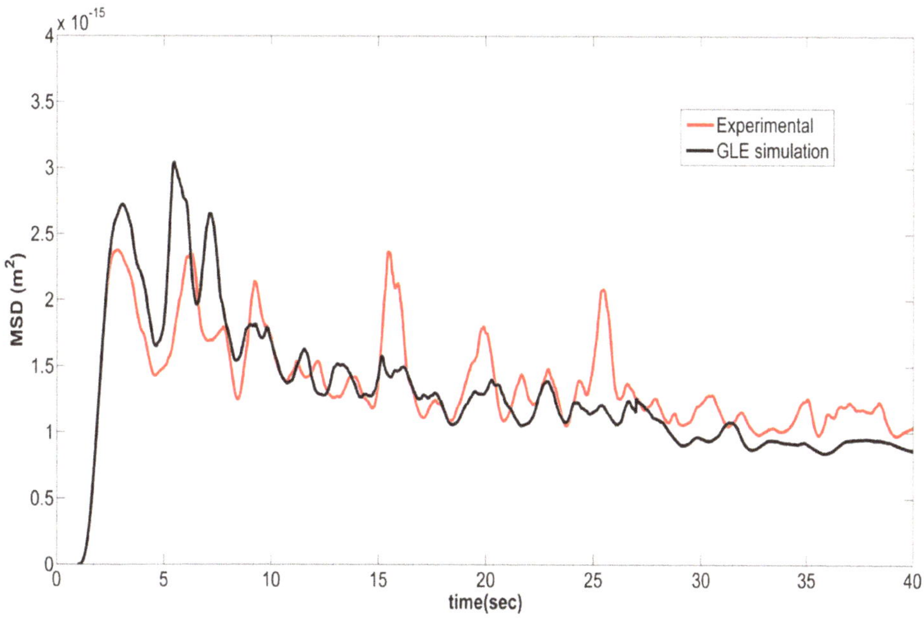

Figure 5.5. Same as figure 5.3 but for slab of pork back-fat.

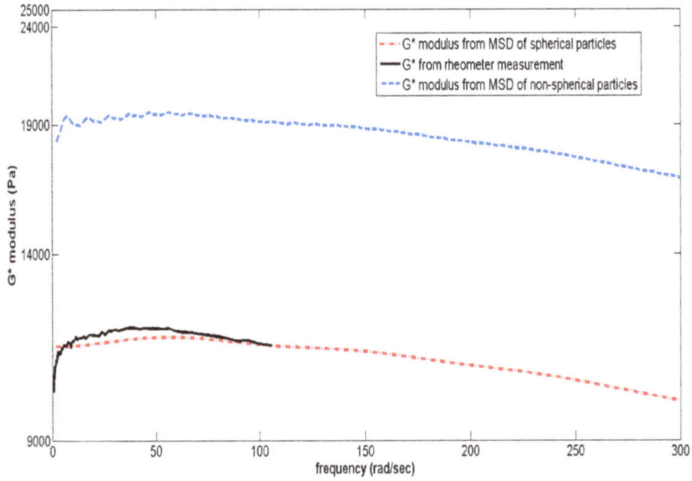

(a)

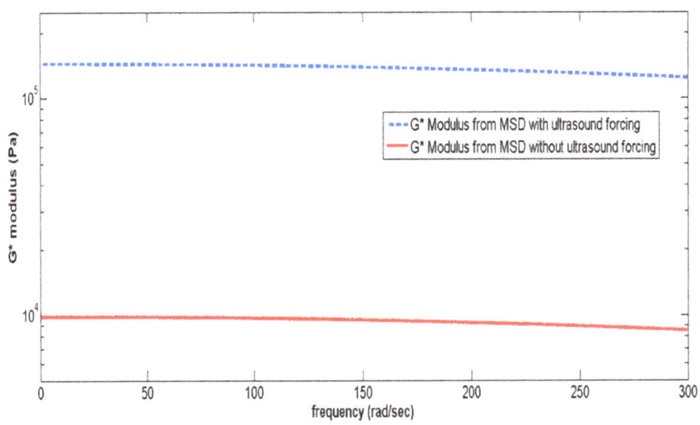

(b)

Figure 5.7. (a) Complex modulus spectrum of elasticity of agar, obtained from DWS measurements: (blue) non-spherical particles and using linear GSER; (red) with spherical particles and linear GSER; (black) using rheometer. (b) Same as (a) but for pork. Here, the scattering centres are anisotropic. Therefore, linear diffusion GSER gave results which are quite removed from the one with rotations. Withdrawing ultrasound we have made the contribution from micro-rotations smaller and the recovered complex modulus spectrum smaller (red curve). Rheometer data is missing as thin samples of pork fat with uniform thickness was difficult to prepare.

within the ROI with a loss of memory of the initial direction; i.e. the volume of the ROI should be small enough.

This discrepancy can be corrected provided we remove from the recovered MSD pertaining to anisotropic scattering the contribution from pure translation, which is the term containing $\langle \Delta L^2(\tau) \rangle$ in equation (5.11). Towards this we make use of the MSD curve recovered from spherical particles, which, in any case, contains only the term from linear displacement. The experiments done with the first agar slab which has spherical particles of the same concentration per unit volume as the non-

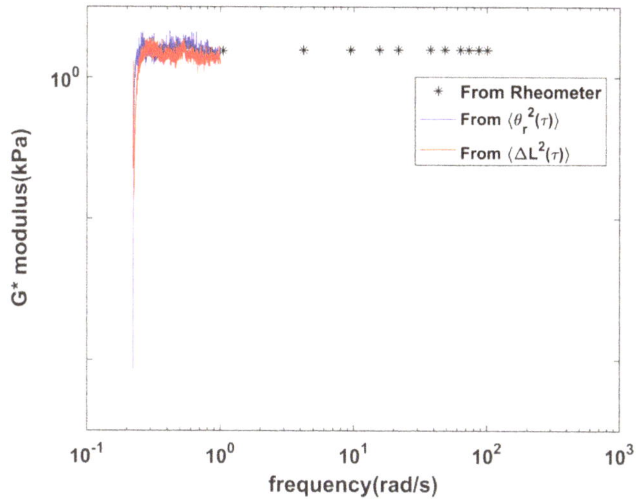

Figure 5.8. Complex modulus spectrum obtained from agar with non-spherical scattering particles. Rotational contribution to phase fluctuation is separated out using experimental data from spherical particles. Since agar is isotropic the complex modulus spectra from rotational diffusion (blue) matches that from translational one (red). Both are verified by independent measurement from rheometer (**).

spherical ones, gets us the corresponding $M(\tau)$ and also $\langle\Delta L^2(\tau)\rangle$. Subtracting out this from the combined term in equation (5.11) we get the term containing $\langle\theta_r^2(t)\rangle$. Employing this and using the rotational diffusion GSER we recover $|G^*(\omega)|$ (from equation (2) of [1]). This is plotted in figure 5.8, which closely matches the spectrum obtained from the linear-diffusion GSER. This match is owing to the fact that agar is a mechanically isotropic material.

5.6 Conclusion

In this work, we have demonstrated rotational-cum-translational diffusion micro-rheology with the help of light diffusion from an ensemble of anisotropic scattering centres in, primarily an agar phantom. In comparison to light scattering from similar sized spherical particles, phase fluctuation picked-up from particles with anisotropy is larger owing to contributions from micro-rotations. This larger phase fluctuation affects the correlation decay rate, which, in turn, affects the MSD evolution, both in its transient and the noise in the plateau region. The increased slope in the transient scales the reconstructed complex modulus spectrum. With an additional set of experiments under similar conditions with the same material with isotropic scatterers, we have recovered MSD variation from pure translatory motion, which was used to extract the rise in mean-square rotation with time. From this, using GSER for rotational diffusion we have been able to recover complex modulus of elasticity matching the one obtained from translational diffusion. In the experimental demonstration, we have used focused ultrasound to both delineate a region in the object being studied and to excite micro-rotation of the scattering centres. Thus we have been able to measure the effect of rotational diffusion of Brownian particles using DWS measurements. This would open new avenues to study cross-over from anisotropic particle transport to isotropic diffusion, as discussed in [2]. The major reason for the possibility of such measurement is the isolation of a small enough region in the object where anisotropic particle diffusion can be observed by working with anisotropic light scattering.

It is also seen that the intensity of noise in the plateau is connected to the particle shape anisotropy and can be used as a measure to characterize the morphology of the scattering centres, with potential biomedical diagnostic applications. The dynamics of the scattering centres is modelled using the GLE with an internal, multiplicative noise term capturing micro-rotations. With this model as the 'forward model' we are able to recover both the amplitude of the internal noise representing anisotropy of particles, and the Prony series coefficients representing the average size of them. To the best of our knowledge, this reports the first experimental measurement of rotational diffusion through light-diffusion studies through particles with shape anisotropy.

References

[1] Cheng Z and Mason T G 2003 Rotational-diffusion micro-rheololgy *Phys. Rev. Lett.* **90** 018304

[2] Han Y *et al* 2006 Brownian motion of an ellipsoid *Science* **314** 626–30

[3] Weitz D A and Pine D J 1993 *Diffusing-Wave Spectroscopy Dynamic Light Scattering: The Method and Some Applications* ed W Brown (Oxford: Oxford University Press)

[4] Dasgupta B R and Weitz D A 2005 Microrheology of cross-linked polyacrylamide networks *Phys. Rev.* E **71** 021504

[5] Sarkar S *et al* 2015 Internal-noise driven generalized Langevin equation for a nonlocal continuum model *Phys. Rev.* E **92** 022150

[6] Baczewski A D and Bond S D 2013 Numerical integration of an extended-variable generalized Langevin equation with a positive Prony representable memory kernel *J. Chem. Phys.* **139** 044107

[7] Roy D and Rao G V 2017 *Stochastic Dynamics, Filtering and Optimization* (Cambridge: Cambridge University Press)

[8] Venugopal M *et al* 2016 Quantitative photo-acoustic tomography by stochastic search: direct recovery of optical absorption field *Opt. Lett.* **41** 4202–5

Ram Mohan Vasu and Debasish Roy

Chapter 6

Concluding remarks

The main objective of the present study is to throw light on the overall dynamics of the object at its ultrasound irradiated region, the ROI. It is this dynamics, which causes the coherent light interrogating the object to suffer a phase fluctuation which in turn is the major reason for the detected speckle-intensity modulation, the measurement. The other reason is the optical contrast, primarily the optical absorption coefficient ($\mu_a(\mathbf{r})$) in the region and also the scattering coefficient, which limits the number of photons coming out of the region. We first brought home the necessity of accounting for *all* the material properties, optical as well as mechanical, which influence the measurement of UMOT, the speckle modulation depth, in modelling the measurement and quantitatively accurate recovery of the optical contrast from it. This is important in the medical diagnostic imaging context, for malignancy results in both absorption coefficient and mechanical stiffness increase in its wake. We have only concentrated on elastic property recovery; all the same, it will be interesting to see how recovered $\mu_a(\mathbf{r})$ values change after effect of shear modulus change in the ROI on speckle-modulation depth is properly accounted for. Equally, it will also be interesting to see how shear modulus recovered would be altered, if $\mu_a(\mathbf{r})$ changes are accounted for. One way out of the difficulty, if variation in one set of properties is unknown, is to use a measurement which is insensitive to that set of properties. Since we are concerned only with elastic property recovery, we have employed certain measurements which are insensitive to optical properties. The property we have used is mainly the vibrational spectral modes of the ROI, or in a fuller measure, vibrational spectrum of the ROI. For $\mu_a(\mathbf{r})$ recovery we could not settle on a satisfactory measurement which is on one hand quite sensitive to $\mu_a(\mathbf{r})$ but insensitive to elastic property. Perhaps a measurement derived from the detected speckle modulation, for example, integral of the square of the modulation, can answer this lack, which could be low sensitivity to $\mu_a(\mathbf{r})$ in the ROI, but quite insensitive to stiffness.

doi:10.1088/2053-2563/aae893ch6

The property of light which is sensitive to dynamics in the object is its temporal coherence. This is quite profitably used in DWS experiments. There, the Brownian motion of scattering centres causes decay in its field autocorrelation, which is measured and made use of to get the temporal evolution of MSD of Brownian particles, and from it the host material's visco-elastic spectrum. In our first experiment we have extended DWS to probe objects with spatially inhomogeneous visco-elastic properties. For this we used 'remote palpation' from focused and intersecting ultrasound beams. The signature of the ultrasound beam, which here is the presence of a low-frequency acoustic radiation and the consequent sinusoidal movement superimposed on the Brownian particles in the ROI produces also a superimposition on the output autocorrelation: a periodic modulation on the decaying autocorrelation. The temporal decay of the modulation carries information on the Brownian particles within the ROI. From this we recovered the local MSD and visco-elastic spectrum. Since flow of turbid liquid in a capillary within the object presents itself as a dynamic inhomogeneity, we could detect and take steps to measure the flow in the capillary.

That modulation depth can be employed to detect the vibrational modes of the ROI, excited by sinusoidal force from the ultrasound, is demonstrated in chapter 3. The strategy is to scan the (low) acoustic frequency of the heterodyne signal from ultrasound beams. The modulation depth varies and attains a peak at the natural frequency of the VROI. Unfortunately, the higher order modes, being weak, could not be detected by this method. In the second part of the same chapter, we measure, not only the peak to ascertain the natural *frequency* of the VROI, but the spectral variation in full of the modulation depth, representing the vibrational spectrum of the ROI. This enriched data set is used to tomographically recover Young's modulus distribution of the material of the ROI corresponding to an inhomogeneous distribution of this property. Even though modulation depth is dependent also on local $\mu_a(\mathbf{r})$, its acoustic frequency dependence is not. Therefore, in common with our modal frequency data, the spectral variation of modulation can also be considered independent of $\mu_a(\mathbf{r})$.

Driven by ultrasound force, the ROI is subjected to vibration, creating a compressional (acoustic) wave. This compressional wave was well researched in the past and is known as the vibro-acoustic (VA) wave. In our strategy towards quantitative shear modulus recovery by detecting the VA wave on the surface of an object, we have devised a model to explain how shear-wave information at the ROI modulates the VA wave and is carried to the detector at the surface. This is discussed in chapter 4, where we have detected the VA wave and used the spectral peak thereof to detect the natural frequency of vibration of the ROI. We have used a novel detector for VA wave, the fibre Bragg grating (FBG) sensor. Here again, since the higher order modes are weak we could hardly discern their presence from the signal detected through the FBG. It was felt that the higher order modes needed amplification, for making them strong enough to be noticed and detected. It was a chance discovery that an amplified version of a number of modes were contained in the fluctuations (considered a noise before) seen in the plateau of the MSD versus time-delay plot obtained from an ultrasound-assisted DWS experiment targeting the

ROI. With a single modal frequency from the FBG-based measurement we could recover a single shear modulus of the material of the ROI, which was isotropic and homogeneous. But with enriched data from the MSD plateau we could recover six unknowns (out of the nine) from pork fat which is assumed to have orthotropic behaviour. The amplification is explained using stochastic resonance (SR) and array-enhanced SR (AESR) which involve resonant energy transfer from noise background to weak periodic signals (natural modes of the ROI) in a nonlinear system. A word about noise intensity tuning to achieve SR. Additive noise has its origin here in Brownian motion and the intensity of the Brownian noise in an experiment can be controlled by temperature. But to vary temperature is cumbersome; but as mentioned in chapter 4, the background high-frequency acoustic pressure variation owing to mixing of the ultrasound beams can serve as noise, whose intensity can be controlled by selecting the intensity of one of the ultrasound beams. Multiplicative noise is simply due to the visco-elastic connectivity of the scattering particles provided by the host material and therefore can be 'tuned' only by preparing different samples whose interconnection strength is different by changing the storage modulus. We reiterate that such tuning to resonance was not done in the experiments reported here. This certainly can be done, which is left for future implementation. At resonance, we believe with amplification at its highest, many more modes will become visible for measurement of frequency, quite helpful when dealing with truly anisotropic material.

Finally, we have reported, for the first time to the best of our knowledge, experimental measurement of rotational diffusion. We made use of scattering particles with shape anisotropy and the confinement offered by the intersecting ultrasound focal volumes forming the ROI to sense and measure phase fluctuations owing to rotational diffusion of scattering centres. What is not done here, which we believe should be done to prove the point, is to show that as the volume of the ROI increases, the 'memory' of the initial direction of light entry into the ROI is lost due to the onset of scattering isotropy in larger volumes. However, we have shown that the complex modulus of elasticity recovered from the measurement of mean-squared rotation is identical to that obtained from MSD obtained from pure translatory motion. We have also shown that the phase fluctuation measured with anisotropic scattering centres has contributions both from linear movement and rotation. When, naively, GSER for translational diffusion is applied to this data, the recovered complex modulus of elasticity does not match that obtained using rheometer measurements. The present experiments will open avenues for exploring particle diffusion in soft condensed matter materials, particularly to study mixing of these over time and space. In a medium with anisotropic material properties, the information provided by visco-elastic spectrum from rotational diffusion will be different from, and complementary to the information in similar measurement from linear diffusion.

Appendix A

Internal noise driven generalized Langevin equation to model the dynamics of scattering centres in ultrasound focal volume

In chapter 2 we saw that the steady-state mean-square displacement (MSD) of Brownian particles in the ultrasound focal volume shows large-scale fluctuations. In addition, it is also noticed that these fluctuations disappear when the ultrasound forcing is switched off. In order to study the response of a material such as the PVA, whose internal length scale is comparable to that of the forcing, using a continuum material model with force-balance equation, one has to take into account also the material microstructure. This renders the classical continuum hypothesis of strictly local interactions invalid and the response model should also accommodate non-local interactions. Within the continuum hypothesis there are non-local modelling techniques such as micropolar [1], micromorphic [1] and gradient theories [2] incorporating long-range inter-particle interactions. In materials such as polymers, granular solids and others with length scales of macroscopic order, long-range effects could predominate, and in such cases predictions through non-local models are in closer conformity to experimental observations. In one of the modelling approaches, the continuum model itself is replaced by a collection of harmonic oscillators with emphasis on a single oscillator (or particle) representing the motion of a small region, and one arrives at a generalized Langevin equation (GLE) for the predominant translational degree of freedom (DOF) of the particle, after incorporating the coupling effects of the neighbouring particles. Even though the GLE could be a handy oscillator based representation for the infinite dimensional continuum, in its standard form it does not include length scale information as is done in non-local continuum theories. Since the standard GLE does not make provision for relevant micro-structural effects in the material, simulations using this failed to reproduce

doi:10.1088/2053-2563/aae893ch7

experimentally observed large-scale fluctuations in the steady-state MSD from the PVA object. In this work we have used a modified form of the GLE, which carries microstructural information and in the present appendix we derive a modified GLE which captures also the long-range microstructural effects of the object. This also helps us in understanding the so called mixing of stochastic and deterministic dynamics as manifested in the large-scale fluctuations in the MSD plateau.

What is new in the modified form of the GLE is the appearance of an internal noise term whose origin is modelled arising out of micro-rotations suffered by the particles owing to ultrasound forcing (in addition to translation). The noise is a consequence of the uncertainty involving two strain operators of which only one contains micro-rotational information. Through this internal noise, the GLE models evolution of the microstructural interactions, similar to a full-blown non-local continuum theory.

A.1 Derivation of the discrete Hamiltonian incorporating micropolar rotation

Our fundamental assumption is that the adoption of micropolarity is enough to capture the microstructural information. Accordingly, following the micropolar continuum theory, the deformation kinematics requires each material point to have microrotation DOFs in addition to the usual translational DOFs. The overall mechanical energy functional (Π) for a geometrically non-linear micropolar body of material volume V_0 may be represented as $\Pi = \frac{1}{2}\int_{V_0} \rho_0(\mathbf{v}\cdot\mathbf{v} + J\boldsymbol{\omega}\cdot\boldsymbol{\omega})d\mathbf{V} + \int_{V_0} \rho_0\psi d\mathbf{V}$. Here $\mathbf{v}$ is the linear velocity and $\boldsymbol{\omega}$ the spin velocity, ρ_0 the mass density, $\rho_0 J\mathbf{I}$ the micro-inertia tensor and ψ the specific free-energy potential, all represented in Lagrangian coordinates. Assuming micro-rotations to be small, a consistent spatial discretization of Π leads to the Hamiltonian (H) corresponding to a discrete representation of the body. In the discrete body, surrounded by a set of bath particles is a system particle whose predominant translational DOF, also called the system DOF, is described by the GLE we derive here. Except for the system DOF, all the other DOFs appearing in H are referred to as the bath DOFs. Accordingly, we split the Hamiltonian as $H = H_s + H_b$, where $H_s = \dfrac{p_{11}^2}{2m_{11}} + w_{1111}u_{11}^2$ does not involve bath DOFs (see [3] for elaboration and further details of the symbols used) and

$$
\begin{aligned}
H_b = {} & \frac{p_{21}^2}{2m_{21}} + \frac{p_{31}^2}{2m_{31}} + \sum_{\alpha=2}^{N}\frac{p_{j\alpha}^2}{2m_{j\alpha}} + \sum_{\alpha=2}^{N}\frac{p_{\theta j\alpha}^2}{2I_{j\alpha}} \\
& + \sum_{\alpha,\beta=2}^{N} w_{ji\alpha\beta}u_{j\alpha}u_{i\beta} + \sum_{\beta=2}^{N} w_{i11\beta}u_{11}u_{i\beta} \\
& + \sum_{\alpha,\gamma=1}^{N} \tilde{w}_{i\alpha\gamma}u_{j\alpha}\{(\theta_{j\gamma}\theta_{i\gamma} + 2S(\theta_{k\gamma})_{k\neq j,\, i})_{i\neq j}\} \\
& + \sum_{\alpha,\gamma=1}^{N} \tilde{w}_{i\alpha\gamma}u_{j\alpha}\left(1 - \sum_{k\neq i}\theta_{k\gamma}^2\right)_{i=j}
\end{aligned}
$$

Here $i, j \in \{1, 2, 3\}$ are the indices for the three Cartesian coordinates. Summations over the repeated indices (i, j) are implied. $m_{k\alpha}$ and $I_{k\alpha}$ are, respectively, the mass and mass-moment of inertia of the αth particle. The coordinate index k, fictitiously introduced in m and I, serves to maintain indicial consistency with the vectors $\mathbf{u}$, $\boldsymbol{\theta}$ and $\mathbf{p}$ appearing in the expressions and helps in using Einstein's summation convention in k. $p_{i\alpha}$, $p_{\theta i\alpha}$ are, respectively, the linear and angular momentum components and $u_{i\alpha}$, $\theta_{i\alpha}$ the displacement and micro-rotation components, all evaluated at the αth particle in the ith direction. Of specific interest are u_{11} and p_{11}, the displacement and the momentum components of the system particle in the required direction, for which the GLE will be written. For $(j, i) = (1, 2), (2, 3), (3, 1)$, $S = 1$ and for other combinations of (j, i), $S = -1$. N is the number of particles in the discrete body.

A.2 Formulation of the new GLE

Using the discrete Hamiltonian H, the governing dynamics for the system and bath variables, described in terms of the displacement DOFs and the momenta, are obtained through Hamilton's equations; details of the derivation are provided in [3]. The equations of motion, in a compact form, are given by equations (A.1) and (A.2).

$$\frac{d}{dt}\begin{Bmatrix} u_{11} \\ p_{11} \end{Bmatrix} = \begin{pmatrix} 0 & 1/m_{11} \\ \hat{w}_{1111} & 0 \end{pmatrix}\begin{Bmatrix} u_{11} \\ p_{11} \end{Bmatrix} + \boldsymbol{\Upsilon}\begin{Bmatrix} \hat{u}_{\theta} \\ \hat{p}_{u\theta} \end{Bmatrix} \tag{A.1}$$

where

$$\boldsymbol{\Upsilon} = \begin{pmatrix} \mathbf{0}_{1\times(6N-1)} & \mathbf{0}_{1\times(6N-1)} \\ \mathbf{w}_{1\times(6N-1)} & \mathbf{0}_{1\times(6N-1)} \end{pmatrix} \text{ and}$$
$$\frac{d}{dt}\begin{Bmatrix} \hat{u}_{\theta} \\ \hat{p}_{u\theta} \end{Bmatrix} = \begin{pmatrix} \mathbf{0}_{(6N-1)\times(6N-1)} & \mathbf{A} \\ \mathbf{K} & \mathbf{0}_{(6N-1)\times(6N-1)} \end{pmatrix}\begin{Bmatrix} \hat{u}_{\theta} \\ \hat{p}_{u\theta} \end{Bmatrix} + u_{11}\begin{Bmatrix} \mathbf{0}_{(6N-1)\times 1} \\ \mathbf{g} \end{Bmatrix} \tag{A.2}$$

$\hat{u}_{\theta} = (\hat{u}^T \ \ \theta^T)^T = \{u_{21} \ \ldots \ u_{3N} \ \ \theta_{11} \ \ldots \ \theta_{3N}\}^T$ consists of displacements and micro-rotations of the bath variables and $\hat{p}_{u\theta} = \{p_{21} \ \cdots \ p_{3N} \ \ p_{\theta 11} \ \cdots \ p_{\theta 3N}\}^T$, the corresponding linear and angular momenta. $\mathbf{A}$ is a diagonal matrix with nonzero entries $\{1/m_{21} \ \ldots \ 1/m_{3N} \ \ 1/I_{11} \ \ldots \ 1/I_{3N}\}$. $\mathbf{K} = (\mathbf{K}_1 \ \ \mathbf{K}_2 + u_{11}\mathbf{K}_3)$, $\mathbf{K}_2 = \begin{pmatrix} \mathbf{A} \\ \mathbf{0}_{3N\times 3N} \end{pmatrix}$ and $\mathbf{K}_3 = \begin{pmatrix} \mathbf{0}_{(3N-1)\times 3N} \\ \mathbf{B} \end{pmatrix}$, $\mathbf{K}_1$, $\mathbf{A}$ and $\mathbf{B}$ are constant matrices of dimensions $(6N-1)\times(3N-1)$, $(3N-1)\times 3N$ and $3N\times 3N$, respectively. g is a $(6N-1)$ dimensional constant vector. $\mathbf{0}_{m\times n}$ designates a zero matrix of dimension $m\times n$ and $\mathbf{w}_{1\times(6N-1)}$ in $\boldsymbol{\Upsilon}$ is a constant matrix. Denoting $\{\hat{u}^T{}_{\theta} \ \ \hat{p}^T{}_{u\theta}\}^T$ as Y_b, equation (A.2) may be written as:

$$\frac{dY_b}{dt} = \bar{\mathbf{K}}Y_b + u_{11}\left\{0_{(6N-1)\times(6N-1)} \ \ \mathbf{g} + \mathbf{B}\theta\right\}^T \tag{A.3}$$

where

$$\overline{\mathbf{K}} = \begin{pmatrix} \mathbf{0}_{(6N-1)\times(6N-1)} & \mathbf{\Lambda} \\ (\mathbf{K}_1 \ \ \mathbf{K}_2) & \mathbf{0}_{(6N-1)\times(6N-1)} \end{pmatrix}$$

Multiplying equation (A.3) with $\exp(-\overline{\mathbf{K}}t)$ and integrating over [0, t], we get the following implicit expression:

$$Y_b(t) = \exp(\overline{\mathbf{K}}t)\, Y_b(0) + \exp(\overline{\mathbf{K}}t)\int_0^t \exp(-\overline{\mathbf{K}}s)u_{11}(s)\left\{\mathbf{0}_{(6N-1)\times 1} \ \ \mathbf{g} + \mathbf{B}\theta(s)\right\}^T ds \quad \text{(A.4)}$$

Integration by parts on the term $\int_0^t \exp(-\overline{\mathbf{K}}s)u_{11}(s)\{\mathbf{0}_{(6N-1)\times 1} \ \ \mathbf{g}\}^T ds$ of equation (A.4) would lead to an equivalent representation given in equation (A.5)

$$\begin{aligned} Y_b(t) = {} & \exp(\overline{\mathbf{K}}t)\, Y_b(0) + \overline{\mathbf{K}}^{-1}u_{11}(t)\left\{\mathbf{0}_{(6N-1)\times 1} \ \ \mathbf{g}\right\}^T \\ & - \exp(\overline{\mathbf{K}}t)\overline{\mathbf{K}}^{-1}u_{11}(0)\left\{\mathbf{0}_{(6N-1)\times 1} \ \ \mathbf{g}\right\}^T \\ & + \exp(\overline{\mathbf{K}}t)\int_0^t \exp(-\overline{\mathbf{K}}s)\overline{\mathbf{K}}^{-1}\dot{u}_{11}(s)\left\{\mathbf{0}_{(6N-1)\times 1} \ \ \mathbf{g}\right\}^T ds \\ & + \exp(\overline{\mathbf{K}}t)\int_0^t \exp(-\overline{\mathbf{K}}s)\overline{\mathbf{K}}^{-1}u_{11}(s)\left\{\mathbf{0}_{(6N-1)\times 1} \ \ \mathbf{B}\theta(s)\right\}^T ds \end{aligned} \quad \text{(A.5)}$$

Substituting $Y_b(t)$, as in equation (A.5), into equation (A.1), we obtain the governing equations of motion (equation (A.6)) for the system variables with the micro-rotation DOFs still tagged on

$$\begin{aligned} \frac{d}{dt}\begin{Bmatrix} u_{11} \\ p_{11} \end{Bmatrix} = {} & \begin{pmatrix} 0 & 1/m_{11} \\ (\hat{w}_{1111} + \breve{w}_0) & 0 \end{pmatrix}\begin{Bmatrix} u_{11} \\ p_{11} \end{Bmatrix} \\ & + \mathbf{\Upsilon}\int_0^t \exp(-\overline{\mathbf{K}}(s-t))\overline{\mathbf{K}}^{-1}\dot{u}_{11}(s)\begin{Bmatrix} \mathbf{0}_{(6N-1)\times 1} \\ \mathbf{g} \end{Bmatrix} ds \\ & + \mathbf{\Upsilon}\left\{\exp(\overline{\mathbf{K}}t)\left(Y_b(0) - \overline{\mathbf{K}}^{-1}u_{11}(0)\begin{Bmatrix} \mathbf{0}_{(6N-1)\times 1} \\ \mathbf{g} \end{Bmatrix}\right)\right\} \\ & + \mathbf{\Upsilon}\int_0^t \exp(-\overline{\mathbf{K}}(s-t))u_{11}(s)\begin{Bmatrix} \mathbf{0}_{(6N-1)\times 1} \\ \mathbf{B}\theta(s) \end{Bmatrix} ds \end{aligned} \quad \text{(A.6)}$$

The constant element $\breve{w}_0$ in equation (A.6) additionally contributes to the stiffness due to micropolarity. This constant is the second element of the two-dimensional vector $\mathbf{\Upsilon}\overline{\mathbf{K}}^{-1}(\mathbf{0}_{(6N-1)\times 1} \ \ \mathbf{g})^T$.

Our interest is in deriving a GLE for the system DOF u_{11}, which requires eliminating the micro-rotation DOFs from equation (A.6). Inherent configurational uncertainty of the microstructure and its time evolution would allow that the micro-rotation DOFs are treated as stochastic processes. A further justification of this viewpoint is provided through the noncommutativity of the symmetrized polar and nonpolar operators, $\hat{\mathbf{E}}_p = \frac{1}{2}(\tilde{\mathbf{E}}_p + (\tilde{\mathbf{E}}_p)^T)$ and $\hat{\mathbf{E}}_{np} = \frac{1}{2}(\tilde{\mathbf{E}}_{np} + (\tilde{\mathbf{E}}_{np})^T)$, wherein $\hat{\mathbf{E}}_p = \mathbf{R}^T\mathbf{F}$ and $\hat{\mathbf{E}}_{np} = \mathbf{F}^T\mathbf{F}$ are, respectively, considered to be measures of polar and nonpolar strains. **R** and **F** are micro-rotation and deformation gradient tensors respectively [3]. Specifically, we can arrive at the following Robertson–Schrodinger uncertainty [4] relation involving these two operators,

$$\sigma_{\hat{\mathbf{E}}_p}\sigma_{\hat{\mathbf{E}}_{np}} \geqslant \frac{1}{4}\left(\left|\left\langle\left[\hat{\mathbf{E}}_p, \hat{\mathbf{E}}_{np}\right]\right\rangle\right|^2 + \left|\left\langle\left\{\hat{\mathbf{E}}_p - \left\langle\hat{\mathbf{E}}_p\right\rangle, \hat{\mathbf{E}}_{np} - \left\langle\hat{\mathbf{E}}_{np}\right\rangle\right\}\right\rangle\right|^2\right) \tag{A.7}$$

[**A**, **B**] is the commutator and {**A**, **B**} the anticommutator of the operators **A** and **B**. For a given symmetric matrix operator **C**, $\langle \mathbf{C}\rangle = \int_\Omega \mathbf{f}(\mathbf{X};\, \mathbf{X}')^T\mathbf{C}(\mathbf{X}')\mathbf{f}(\mathbf{X};\, \mathbf{X}')d\mathbf{X}'$ defines the mean of **C** with respect to **f**. **f**(**X**; •) is a compactly supported vector valued function of $\mathbf{X}'$ with an arbitrary support containing **X** and must be normalized, i.e. $||\mathbf{f}||^2 = \int_\Omega \mathbf{f}^T\mathbf{f}d\mathbf{X}' = 1$. Thus, $\mathbf{f}^T\mathbf{f}$ is interpretable as the density associated with a probability measure. Here, for instance, we choose $\mathbf{f} = \dfrac{\mathbf{X} - \mathbf{X}'}{||\mathbf{X} - \mathbf{X}'||}$, a normalized line segment. Such a mean, though a scalar quantity, should be interpreted as an operator wherever appropriate. $\sigma_\mathbf{A} = \langle(\mathbf{A} - \langle\mathbf{A}\rangle)^T\ (\mathbf{A} - \langle\mathbf{A}\rangle)\rangle$ is the variance (an uncertainty measure) associated with **A**. In terms of **F** and **R**, the inequality (A.7) may be recast as,

$$\sigma_{\hat{\mathbf{E}}_p}\sigma_{\hat{\mathbf{E}}_{np}} \geqslant \frac{1}{4}\left|\left\langle(\mathbf{R}^T\mathbf{F} + \mathbf{F}^T\mathbf{R})\ \left(\mathbf{F}^T\mathbf{F} - \langle\mathbf{F}^T\mathbf{F}\rangle\right)\right\rangle\right|^2 \tag{A.8}$$

The uncertainty relation is nontrivial only if the RHS of the inequality (A.8) is strictly greater than zero and this is indeed the case in general. The RHS of (A.8) can be zero only if $\mathbf{f}^T\mathbf{PQf} = 0$ where $\mathbf{P} = \mathbf{R}^\mathrm{T}\mathbf{F} + \mathbf{F}^\mathrm{T}\mathbf{R}$ and $\mathbf{Q} = \mathbf{F}^\mathrm{T}\mathbf{F} - \langle\mathbf{F}^\mathrm{T}\mathbf{F}\rangle$. This result is obtained upon localization of the integral in the definition of ⟨**PQ**⟩, given Ω is arbitrary. Since **P** is positive definite and **Q** generally nonsingular, **PQ** is nonsingular too and thus $\mathbf{PQf} \neq 0$. Clearly **PQ** is not skew symmetric and we may discount, almost surely, the other possibilities of $\mathbf{PQf} \perp \mathbf{f}$. This ensures, with probability 1, that the uncertainty relation is nontrivial. In other words, while eliminating the micro-rotation DOFs, they should be treated as stochastic processes.

Since, at a time instant, the mean of a micro-rotation DOF would typically be an order smaller than its translational counterpart, we may approximately identify the micro-rotation DOF as a zero mean random variable. Retaining such noise terms in the final GLE is then crucial, as nonlocality necessarily implies nondeterminism [3], an aspect generally overlooked in nonlocal continuum theories. The randomness in the micro-rotations, consequent upon the uncertainty relation equation (A.7), may be contrasted with that in the initial conditions due to thermal fluctuations, yielding an additive noise term in the GLE. This leads us to recast equation (A.6) and write the GLE including the two noise sources as:

$$\begin{aligned}\frac{d}{dt}\begin{Bmatrix} u_{11} \\ p_{11}\end{Bmatrix} &= \begin{pmatrix} 0 & 1/m_{11} \\ (\hat{w}_{1111} + \breve{w}_0) & 0\end{pmatrix}\begin{Bmatrix} u_{11} \\ p_{11}\end{Bmatrix} \\ &+ \int_0^t \begin{Bmatrix} 0 \\ \eta(s - t)\dot{u}_{11}(s)\end{Bmatrix} ds + \begin{Bmatrix} 0 \\ \xi(t)\end{Bmatrix} + \int_0^t \begin{Bmatrix} 0 \\ W_s^t u_{11}(s)\end{Bmatrix} ds\end{aligned} \tag{A.9}$$

where $\hat{\eta}(\cdot)$ is the memory kernel that may be found from the second term on the right-hand side of equation (A.6). $\xi(t)$ is a linear combination of $u_{11}(0)$ and elements of $Y_b(0)$ at time t. An application of Lyapunov's central limit theorem (CLT) yields $\xi(t)$ to be a zero mean Gaussian random variable at time t. On similar lines, the internal (multiplicative) noise W_s^t, arising from a weighted sum of the micro-rotations as seen from the last term of equation (A.6), may also be characterized as zero mean Gaussian for fixed s and t. We may rewrite the new GLE (including an external forcing term $F(t)$ for completion) in its more familiar second order form.

$$m\ddot{u} + ku + \int_0^t \eta(s-t)\dot{u}(s)ds = F(t) + \xi(t) + \int_0^t W_s^t(s-t)u(s)ds \tag{A.10}$$

In equation (A.10), $m = m_{11}$ denotes the mass, $k = -(\hat{w}_{1111} + \check{w}_0)$ the stiffness and $\eta = -\hat{\eta}$ the damping memory kernel for the system variable. Note that if micropolar effects are not considered, i.e. W_s^T and $\check{w}_0$ are identically zero, and the usual form of GLE is retrieved.

A.3 A fluctuation–dissipation (FD) constraint

In addition to the uncertainty constraint as in equation (A.7), the causality condition may impose a constraint on the new GLE. The latter would be similar to the FD theorem typically associated with the conventional GLE. We designate the integral term on the right-hand side of equation (A.10) by $y(t)$. A general scheme to represent W_s^t could be via a Wiener chaos representation [5]. However, for illustrative purposes, we consider the special form $W_s^t = \exp(\alpha(s-t))\zeta(s)$ (α is some constant and $\zeta(s) \sim N(0, \sigma(s))$) that enables representing $y(t)$ as the Markovian solution of a stochastic differential equation. σ is the intensity of ζ. Thus equation (A.10) is equivalent to the coupled set of equations:

$$m\ddot{u} + ku + \int_0^t \eta(s-t)\dot{u}(s)ds = F(t) + \xi(t) + y(t) \tag{A.11}$$

$$\dot{y}(t) = y(t) + u(t)\zeta(t) \tag{A.12}$$

Under a random change of time $t \to \beta(t)$, $u(t)\zeta(t)$ may represented via a zero mean Brownian motion $\hat{B}(\beta(t))$, where $\beta(t) = \int_0^t (\sigma(s)^2 u(s)^2)ds$. Following the derivation of Kubo's second *FDT*, the constraint equation may now be directly written as

$$\langle (z(0) + \xi(0))(z(t) + \xi(t)) \rangle = \eta(t)k_B T \tag{A.13}$$

where $z(t) = \left(\int_0^t \exp(s-t)\hat{B}(\beta(s))ds\right)$

Some manipulations on equation (A.13) would lead to equation (A.14)

$$\int_0^t \exp(s-t)\langle \xi(0)\hat{B}(\beta(s))\rangle ds = \eta(t)k_B T - \langle \xi(0)\xi(t)\rangle \tag{A.14}$$

We see from equation (A.14) that, as the left-hand side vanishes in absence of micro-polarity, we get back the Kubo's FD theorem.

References

[1] Eringen A 1976 *Continuum Physics Volume IV: Polar and Nonlocal Field Theories* (New York: Academic)

[2] Mindlin R D and Eshel N N 1968 On the first strain-gradient theories of linear elasticity *Int. J. Solids Struct.* **4** 109–24

[3] Sarkar S *et al* 2015 Internal noise-driven generalized Langevin equation from a nonlocal continuum model *Phys. Rev.* E **92** 022150

[4] Mason T G and Weiss D A 1995 Optical measurements of frequency-dependent linear visco-elastic moduli of complex fluids *Phys. Rev. Lett.* **74** 1250–3

[5] Priestley M 1967 Power-spectral analysis of nonstationary random processes *J. Sound Vib.* **6** 86–97

IOP Publishing

Ram Mohan Vasu and Debasish Roy

Appendix B

Reconstruction based on stochastic evolution

In chapter 5, the four resonant modes that were captured using the ultrasound-assisted DWS experiment were inverted to recover the elastic tensor characterizing the probed anisotropic tissue region. The inversion technique based on an evolutionary stochastic method involves defining the unknown nine elastic parameters and the four measured resonant frequencies as stochastic processes in an appropriate underlying probability space. The unknown vector, updated with a deterministic scheme, can possibly end up having a number of solutions rendering the inverse problem to be ill-posed. This calls for an elaborate regularization technique with a view to obtaining a unique (if possible) reconstruction, meeting other *a priori* known constraints. The evolutionary stochastic reconstruction method (a form of the class of Bayesian schemes), having a multimodal structure in the underlying (posterior) probability measure, inherently eliminates the problem arising due to its ill-posed nature. The inversion technique is equipped with an update strategy based on the laws of probability to minimize the measurement-prediction misfit. Besides, a stochastic technique fashioned after the Bayesian update can accommodate the statistical properties of the measurement noise that naturally provides it with an upper hand over its deterministic counterpart.

Teresa *et al* [1] successfully reconstructed a refractive index profile of an optical fibre from the intensity measurements using a similar evolutionary stochastic reconstruction method, which is also used in this work. Following a Bayesian approach, the unknown elastic parameters and the measured modal frequencies, observed to be as random variables are subjected to a prediction-update procedure on their initial realization (also known as ensemble). Also, similar to the method described in [1], this update procedure involves weighing each predicted realization of elastic constants based on the difference between the measured resonant frequencies and those predicted from the model. This form of a recursion technique

doi:10.1088/2053-2563/aae893ch8

clearly assigns greater weights to the realizations with smaller measurement-prediction error, rendering the realisations with smaller weights less likely to be passed into the initial ensemble for the next iteration. This leads to sample degeneracy, a feature that makes the procedure unsuitable in dealing with large dimensional systems.

The purpose of this stochastic evolutionary method is to adjust the multiplicative weight term appropriately to obtain an additive update. This renders the measurement-prediction misfit term with the same statistical properties as the measurement noise, typically a zero-mean random variable of a certain probability distribution. Precisely, we would like to push the misfit to have zero drift eventually as the recursion proceeds, which means that the misfit will be a zero-mean Brownian process, or more appropriately a martingale [2]. This necessarily requires a parameterization of the misfit term (along with other stochastic processes) to evolve over time or (in the absence of quantities which are time varying) pseudotime. Therefore, we introduce a time-like variable τ that strictly increases with iterations: $0 = \tau\ 0 < \tau_1 < \cdots < \tau_k < \cdots$ with τ_k denoting its value at the kth recursion.

The nine independent elastic constants to be recovered along with measured resonant frequencies are described as stochastic processes in τ. Specifically, the stochastic differential equation (SDE) describing the evolution of the parameters to be recovered, and the measurement equation, are, respectively, given by

$$d\mathbf{E}_\tau = d\boldsymbol{\xi}_\tau \tag{B.1a}$$

$$\mathbf{M}_\tau = \mathbf{F}(\mathbf{E}_\tau) + \boldsymbol{\eta}_\tau \tag{B.1b}$$

where $\mathbf{E}_\tau \in \mathbb{R}^9$ denotes the vector whose components represent the nine independent elastic constants to be recovered and $\mathbf{F}$ denotes the measurement operator. It can be noted from equation (B.1a), $\mathbf{E}_\tau$ is a stochastic process in iteration variable τ with $\xi_\tau \in \mathbb{R}^9$ and $\eta_\tau \in \mathbb{R}^4$ being Brownian processes in τ with mean zero and covariance $\sum_\xi \sum_\xi^T \in \mathbb{R}^{9\times 9}$ and $\sum_\eta \sum_\eta^T \in \mathbb{R}^{4\times 4}$, respectively. At the $(k+1)$th iteration, i.e. for the interval $(\tau_k, \tau_{k+1}]$, discretization of equations (B.1a) and (B.1b) yields:

$$\mathbf{E}_{k+1} = \hat{\mathbf{E}}_k + \Delta\xi_k \tag{B.2a}$$

$$\mathbf{M}_{k+1} = \mathbf{F}(\mathbf{E}_{k+1}) + \boldsymbol{\eta}_{k+1} \tag{B.2b}$$

where $(.)_k := (.)_{\tau_k}$, $k \in \mathbb{N}$, $\Delta\xi_k = \xi_{k+1} - \xi_k$ and $\hat{\mathbf{E}}_k$ denotes the last updated estimate of the parameter. Within the Monte Carlo framework used here, the jth sample realization of $\hat{\mathbf{E}}_k$ is denoted as $\hat{\mathbf{E}}_k(j)$. The initial random vectors, $\mathbf{M}_0$ and $\boldsymbol{\eta}_0$, represent, respectively, the experimentally acquired (true) measurement and the (true) measurement noise. It is to be noted that the elements of the covariance matrix $\Sigma_\eta \sum_\eta^T$ of the fictitiously introduced measurement noise η_τ, $\tau > 0$, must be relatively small vis-à-vis $||\mathbf{M}_0||$ in order to ensure that $\mathbf{M}_{k+1}$, $k > 0$ is not very different from (in our case) the experimentally measured modal frequencies, $\mathbf{M}_0$.

After drawing the ensemble of elastic constants, i.e. $\{\hat{\mathbf{E}}_0(j)\}_{j=1}^{n_E}$ (n_E being the ensemble size) from a zero mean Gaussian distribution with a small standard deviation, the predicted set of values at $(k+1)$th iteration (i.e. at $\tau = \tau_{k+1}$) is generated from the last updated ensemble, $\{\hat{\mathbf{E}}_k(j)\}_{j=1}^{n_E}$ at τ_k through equation (B.2*a*). The Monte Carlo version of the prediction equation is given by $\mathbf{E}_{k+1}(j) = \hat{\mathbf{E}}_k(j) + \Delta\xi_k(j), j = 1, \ldots, n_E$. The gain-like correction term that resembles the update of a Kalman filter, as extended for non-linear problems, has been used in this particular work [3]. The update equation can be re-written as

$$\hat{\mathbf{E}}_{k+1}(j) = \mathbf{E}_{k+1}(j) + \mathbf{G}_{k+1}(\mathbf{M}_{k+1} - \mathbf{F}(\mathbf{E}_{k+1}(j))), \quad j = 1, \ldots, n_E. \tag{B.3}$$

Here, $\mathbf{G}_{k+1}$ is the gain matrix that has been computed similar to that of an ensemble square root filter and is given by

$$\mathbf{G}_{k+1} = \mathbf{P}_{k+1}\mathbf{Q}_{k+1}^T\left(\mathbf{Q}_{k+1}\mathbf{Q}_{k+1}^T + \sum\nolimits_\eta \sum\nolimits_\eta^T\right)^{-1} \tag{B.4}$$

where, $\mathbf{P}_{k+1} \in \mathbb{R}^{9\times n_E}$ and $\mathbf{Q}_{k+1} = \mathbb{R}^{4\times n_E}$ are the ensemble perturbation matrices corresponding to the parameters to be recovered, and measurements, respectively. They are expressed as

$$\mathbf{P}_{k+1} = \frac{1}{\sqrt{n_E - 1}} \times \left[\mathbf{E}_{k+1}(1) - \frac{1}{n_E}\sum_{j=1}^{n_E}\mathbf{E}_{k+1}(j), \ldots, \mathbf{E}_{k+1}(n_E) - \frac{1}{n_E}\sum_{j=1}^{n_E}\mathbf{E}_{k+1}(j)\right]$$

and

$$\mathbf{Q}_{k+1} = \frac{1}{\sqrt{n_E - 1}} \times \left[\mathbf{F}(\mathbf{E}_{k+1}(1)) - \frac{1}{n_E}\sum_{j=1}^{n_E}\mathbf{F}(\mathbf{E}_{k+1}(j)), \ldots, \mathbf{F}(\mathbf{E}_{k+1}(n_E)) - \frac{1}{n_E}\sum_{j=1}^{n_E}\mathbf{F}(\mathbf{E}_{k+1}(j))\right]$$

The updated parameters at τ_{k+1} are used in the prediction in equation (B.2*a*) in order to obtain the predictions at the next instant of τ and the recursion is continued. At any iteration, the parameter estimates could be obtained using the empirical mean valueof the random variable, which is,

$$\left\langle \hat{\mathbf{E}}_{k+1} \right\rangle_{n_E} = \frac{1}{n_E}\sum_{j=1}^{n_E}\hat{\mathbf{E}}_{k+1}(j) \tag{B.5}$$

The stopping criterion for a reasonable recovery of the parameters was decided based on the number of iterations it took for the error terms, i.e. the difference between the experimental frequencies and the predicted frequencies for each of the

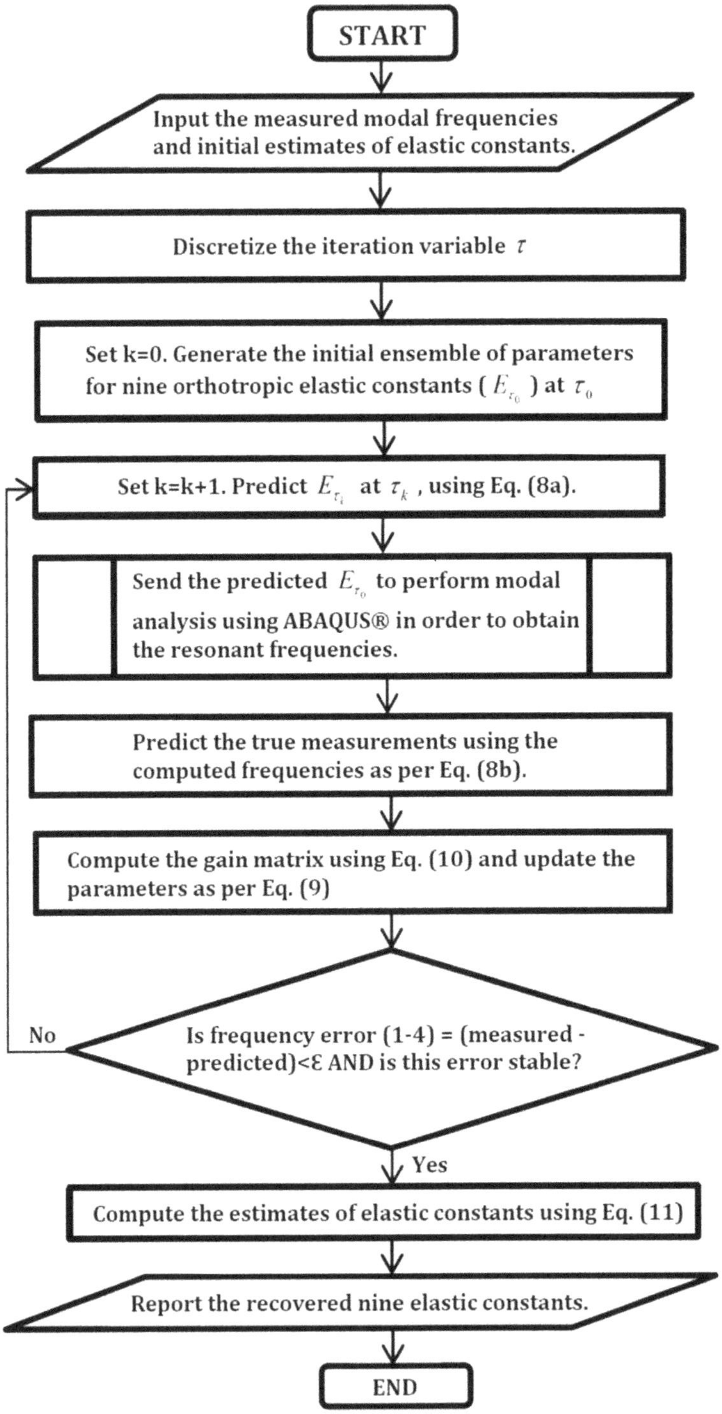

Figure B1. Flowchart explaining the recovery of elastic moduli from the modal frequencies using the evolutionary stochastic search scheme, where ε is a predefined value, used as a stopping criterion.

four modes, to reach individually a steady state below a predetermined tolerance ε. The flowchart explaining the implementation of the evolutionary stochastic method has been shown in figure B1.

References

[1] Teresa J *et al* 2014 Diffraction tomography from intensity measurements: an evolutionary stochastic search to invert experimental data *J. Opt. Soc. Am.* A **31** 996–1006

[2] Klebaner F C 2005 *Introduction to Stochastic Calculus with Applications* vol 57 (Singapore: World Scientific)

[3] Roy D and Rao G V 2017 *Stochastic Dynamics, Filtering and Optimization* (Cambridge: Cambridge University Press)

Lightning Source UK Ltd.
Milton Keynes UK
UKHW051437240319
339674UK00004B/34/P

9 780750 317481